1st International Conference on Big Data and Cloud Computing (ICoBiC 2017)

Journal of Physics: Conference Series Volume 1018

Kuching, Malaysia
25 – 27 November 2017

ISBN: 978-1-5108-6401-6
ISSN: 1742-6588

Printed from e-media with permission by:

Curran Associates, Inc.
57 Morehouse Lane
Red Hook, NY 12571

Some format issues inherent in the e-media version may also appear in this print version.

Printed by Curran Associates, Inc. (2018)

Institute of Physics
Dirac House, Temple Back
Bristol BS1 6BE UK

Phone: 44 1 17 929 7481
Fax: 44 1 17 920 0979

techtracking@iop.org

Additional copies of this publication are available from:

Curran Associates, Inc.
57 Morehouse Lane
Red Hook, NY 12571 USA
Phone: 845-758-0400
Fax: 845-758-2633
Email: curran@proceedings.com
Web: www.proceedings.com

Table of contents

Volume 1018

1st International Conference on Big Data and Cloud Computing (ICoBiC) 2017

25–27 November 2017, Kuching, Sarawak, Malaysia

Accepted papers received: 1 May 2018
Published online: 1 June 2018

Preface

Papers

IOP Conf. Series: Journal of Physics: Conf. Series **1018** (2018) 011001 doi:10.1088/1742-6596/1018/1/011001

Preface/Introduction

The quest for knowledge has become part of national main agenda. In line with the country's aspiration to enhance research and development in the country, UniMAP has been proactively involved and initiating programs and events, which would encourage high impact research encompassing various fields of expertise.

The Centre of Excellence Geopolymer and Green Technology (CEGeoGTech) has been playing a significant role in this quest by taking up this challenge. Thus, one of the initiatives grabbed by the Green Advanced Computing and Technology (GREAT), CEGeoGTech is in organizing this conference itself, which provides a fine platform for knowledge sharing and research ideas exchange in the field of Big Data, Cloud Computing, Green and Sustainable Computing, Technology Science and Management in Sports.

This is indeed an opportune moment to strengthen the bond among researchers and professionals from the academics, the industries and the public sectors in sharing new ideas, and in collaborating future research activities. The importance of interaction and networking among peers are the inspiration of this joint conference. I would like to congratulate the organizing committee and all parties involved for their sheer dedication and tireless effort in making this joint conference a great success. I hope that all participants will have a productive and pleasant conference.

IOP Conf. Series: Journal of Physics: Conf. Series **1018** (2018) 011002 doi:10.1088/1742-6596/1018/1/011002

Peer review statement

All papers published in this volume of *Journal of Physics: Conference Series* have been peer reviewed through processes administered by the proceedings Editors. Reviews were conducted by expert referees to the professional and scientific standards expected of a proceedings journal published by IOP Publishing.

IOP Conf. Series: Journal of Physics: Conf. Series **1018** (2018) 012001 doi:10.1088/1742-6596/1018/1/012001

LiFi based automated shopping assistance application in IoT

Sharmin Akter[#], Dr. Rashidah Funke Olanrewaju*, Thouhedul Islam[#], Salma[#]

Electrical and Computer Engineering, International Islamic University Malaysia
Kuala Lumpur, Malaysia

[1]s_akter226@yahoo.com, [3]tisuchi@gmail.com, [4]mirsalma12@yahoo.com

[*]*Assistant Professor, Electrical and Computer Engineering, International Islamic University Malaysia*

[2]frasidah@iium.edu.my

Abstract. Urban people minimize shopping time in daily life due to time constrain. From that point of view, the concept of supermarket is being popular while consumers can buy different items from same place. However, customer spends hours and hours to find desired items in a large supermarket. In addition, it's also required to be queued during payment at counter that is also time consuming. As a result, a customer has to spend 2-3 hours for shopping in a large superstore. This paper proposes an Internet of Things and Li-Fi based automated application for smart phone and web to find items easily during shopping that can save consumer's time as well as reduce man power in supermarket.

1. Introduction

Modern people are so engaged with multiple daily life activities. Maintaining family, office, business and personal life, sometimes it is very hard for city dwellers to spend time shopping. From that corner, the supermarket is one of the part and parcel in modern life. People prefer shopping from supermarket where cloths, groceries, toiletries, jewelleries and so on items are able to buy from same place [1-2][5]. Since customer buy different items from different corner in a big supermarket, however, it is possible to pay in a one stop counter. Such kind of facilities of a super market has drawn attention of busiest customers [3-4]. As a consequence urban people usually go to supermarket for daily shopping.

While people are busy with their own activities, most of the time, many of them prefer do shopping during weekend. Unlikely during weekend many of them go for shopping, as a result, supermarket are always crowded. In a big supermarket, it is also time consuming to find desire items. In a result, a customer has to spend 1-2 hours to pick their items into trolley. Besides that, there is a long queue during payment while it is crowded. To sum up, the whole process unnecessarily took important hours of customer's daily life.

In this paper, Internet of Things (IoT) and Li-Fi based an automated application has been proposed that helps a customer to find the items from different rows. In addition, it will list down and calculate the price of items and save shopper's time during payment. The next portion of this paper explains the proposed architecture of this application. The following part elaborates the evaluation of proposed work and benchmark the coding structure of this proposed work.

IOP Conf. Series: Journal of Physics: Conf. Series **1018** (2018) 012001 doi:10.1088/1742-6596/1018/1/012001

2. Architecture

This application can proposed in two different versions, web and mobile version. Customer and cashier interface are main two parts of this application's architecture. Figure 1 shows the different modules of this application.

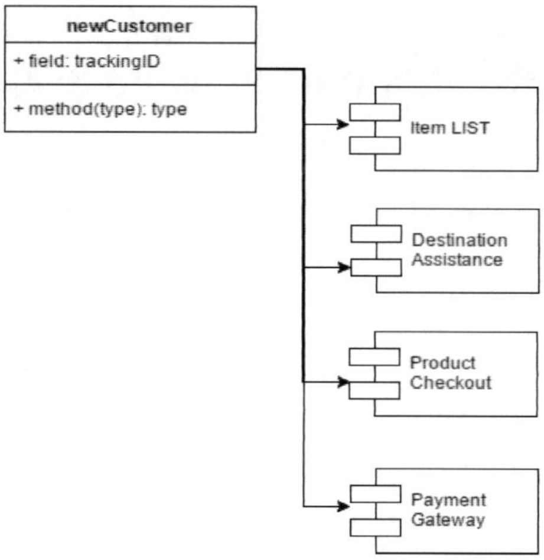

Figure 1. All modules.

2.1 Customer Interface

The customer interface contains few stages. Users are able to create shopping list. Application will assist the list of items by showing row location. Customers are able to get the item's price by scanning barcode using smart phone. At the end of the shopping customer's will get the total price that need to be paid.

Every time a new customer enter in supermarket territory, a free LiFi based internet connection available to connect. The default landing page is the supermarket's online application where user are able to create a new list of item that they need to buy.

2.1.1. *List.* In application, users are able to create a shopping list. Based on user's device IP and MAC address [7], a new list will be generated for a newly connect user that will connect in central database though LiFi and IoT [13]. Based on different user, application automatically create a unique number for each that can be tracking number during payment and reference. Based on availability of product in the super market, users are able to key-in at shopping list. The list of the item will be auto suggested in shopping list that can assist buyer to choose as well. Figure 2 shows the database table design for item list.

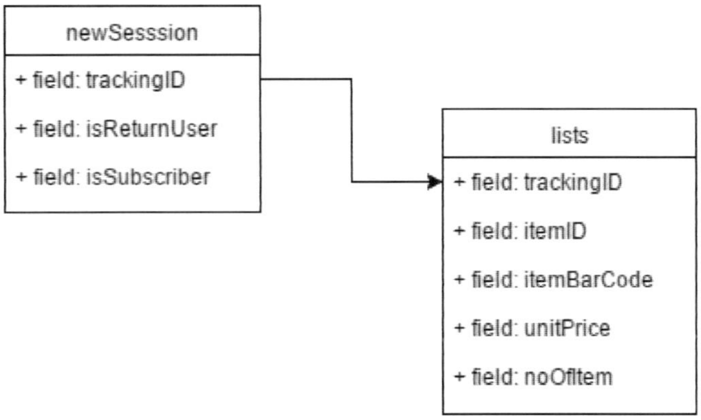

Figure 2. Database design for item list.

2.1.2 Destination Assistance. In application, users are able to create a shopping list. Based on user's device IP and MAC address [7], a new list will be generated for a newly connect user that will connect in central database though LiFi and IoT [13]. Based on different user, application automatically create a unique number for each that can be tracking number during payment and reference. Based on availability of product in the super market, users are able to key-in at shopping list. The list of the item will be auto suggested in shopping list that can assist buyer to choose as well. Figure 2 shows the database table design for item list.

```
(3959 * acos( cos( radians(37) ) * cos( radians( lat ) ) * cos( radians( lng ) - radians(-122) ) + sin( radians(37) ) * sin( radians( lat ) ) ) )
```

Code 1. Code for detecting nearest coordinate.

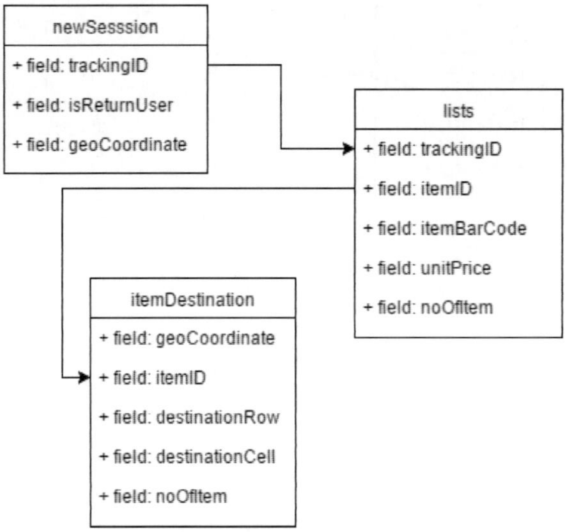

Figure 3. Database design for location distance measurement.

2.1.3 Product Checkout. Users are able to self-checkout the product by using their smart phone. In application, user are able to scan the product barcode and store the barcode number next to the respective item [9]. Based on product's barcode, the price will be added where list will be checked after scanning all the item's barcode. At the end of the procedure, the total price of the shopping will be displayed.

2.2 Cashier Interface

The cashier interface is total control in cashier side. It shows the current active customers who has completed shopping list by scanning barcode. Cashier in-charge can able to search a user by tracking number that was generated during creating a new shopping list at the beginning.

2.2.1 *Paying at Counter.* At the stage of check out from the supermarket, user need to confirm at cashier counter. Cashier in-charge are able to see the list of the current active customer based on nearest location from the cashier's geo location. Apart from that, cashier in-charge are able to find a user by customer's tracking ID. To find the nearest customer, octree algorithm has been used [10-11]. **Code 2** shows the pseudo code of octree algorithm.

```
1. X = Q.X
2. Y = Q.Y
3. Z = Q.Z
4. X = Q.X + cubeSize
5. Y = Q.Y + cubeSize
6. Z = Q.Z + cubeSize
```

Code 2. Pseudo code of octree algorithm.

2.2.2 Payment Gateway. There are 3 types of payment methods has been proposed for this application, card, mobile banking and cash. Cashier helps customers to swap the card or confirm the mobile payment gateway or charge by cash on spot. Based on cashier's confirmation, a bill will be generated and printed for customer from central database with the help of IoT.

3. Design Pattern and Benchmarking

To implement the proposed application, PHP / Python / Java will be used as a code programming language. Beside this, Angular JS, React JS, VueJS, Pusher, Socket.io, Radis will be used for making the application real time.

Since it's a large application and everyday there are lot of customers, the big data is the huge concern for this application. The mysql will be used to store the data. In order to optimise the big data, mysql data driven model will be used [12][14]. The whole application will be stored in a central database from where all the data will be represented with the assist of IoT.

The xhprof will be used in order to benchmark the whole application [11][13]. **Code 3** show the benchmarking pseudo code for testing unit of application performance.

```
> use xhprof
db.results.ensureIndex( { 'meta.SERVER.REQUEST_TIME' : -1 } )
db.results.ensureIndex( { 'profile.main().wt' : -1 } )
db.results.ensureIndex( { 'profile.main().mu' : -1 } )
db.results.ensureIndex( { 'profile.main().cpu' : -1 } )
db.results.ensureIndex( { 'meta.url' : 1 } )
```

Code 3. Benchmarking code by xhprof.

4. Conclusion

Although people are spending money for shopping, however, sometimes it becomes very difficult to spend time for that. In such kind of cases, customers prefer easiest way for shopping. In order to mitigate customer's searching and waiting time in super market, the proposed application can contribute tremendously. This application provide flexibility to customers to choose either web or mobile application for smoothly uses. Since this application assist customers to find out desire shopping item by calculating congesting time of current situation and self-checkout, intimately that can save 2-3 hours of a weekly customer's shopping time.

References

[1] Y. M. Yusoff, R. Rosli, M. U. Karnaluddin, and M. Samad, "Towards smart street lighting system in Malaysia," in 2013 IEEE Symposium on Wireless Technology Applications (ISWTA), 2013, pp. 301–305.

[2] R. Mahendran, "Integrated LiFi(Light Fidelity) for smart communication through illumination," in 2016 International Conference on Advanced Communication Control and Computing Technologies (ICACCCT), 2016, pp. 53–56.

[3] M. Nakagawa, "Microwave and Millimeter-Wave Wireless Home Link System in Japan," in 2000 30th European Microwave Conference, 2000, pp. 1–5.
[4] T. Komine and M. Nakagawa, "Fundamental analysis for visible-light communication system using LED lights," IEEE Transactions on Consumer Electronics, vol. 50, no. 1, pp. 100–107, Feb. 2004.
[5] S. Wu, H. Wang, and C. H. Youn, "Visible light communications for 5G wireless networking systems: from fixed to mobile communications," IEEE Network, vol. 28, no. 6, pp. 41–45, Nov. 2014.
[6] T. Borogovac, M. Rahaim, and J. B. Carruthers, "Spotlighting for visible light communications and illumination," in 2010 IEEE Globecom Workshops, 2010, pp. 1077–1081.
[7] H. Sugiyama, S. Haruyama, and M. Nakagawa, "Brightness Control Methods for Illumination and Visible-Light Communication Systems," in Third International Conference on Wireless and Mobile Communications, 2007. ICWMC '07, 2007, pp. 78–78.
[8] K. Fan, T. Komine, Y. Tanaka, and M. Nakagawa, "The effect of reflection on indoor visible-light communication system utilizing white LEDs," in The 5th International Symposium on Wireless Personal Multimedia Communications, 2002, vol. 2, pp. 611–615 vol.2.
[9] P. P. Salian, S. Prabhu, P. Amin, S. K. Naik, and M. K. Parashuram, "Visible Light Communication," in India Educators' Conference (TIIEC), 2013 Texas Instruments, 2013, pp. 379–383.
[10] Y. Tanaka, T. Komine, S. Haruyama, and M. Nakagawa, "Indoor visible communication utilizing plural white LEDs as lighting," in 12th IEEE International Symposium on Personal, Indoor and Mobile Radio Communications. PIMRC 2001. Proceedings (Cat. No.01TH8598), 2001, vol. 2, p. F–81–F–85 vol.2.
[11] H. Haas, "LiFi: Conceptions, misconceptions and opportunities," in 2016 IEEE Photonics Conference (IPC), 2016, pp. 680–681.
[12] M. Ayyash et al., "Coexistence of WiFi and LiFi toward 5G: concepts, opportunities, and challenges," IEEE Communications Magazine, vol. 54, no. 2, pp. 64–71, Feb. 2016.
[13] Z. Thomas, N. Kumar, and D. J. Preshiya, "Automatic billing system using Li-Fi module," in 2016 International Conference on Communication and Signal Processing (ICCSP), 2016, pp. 2216–2219.
[14] E. Ronen and A. Shamir, "Extended Functionality Attacks on IoT Devices: The Case of Smart Lights," in 2016 IEEE European Symposium on Security and Privacy (EuroS P), 2016, pp. 3–12.

Acknowledgments

Acknowledgments. This work was partially supported by Ministry of Higher Education Malaysia (Kementerian Pendidikan Tinggi) under Research Initiative Grant Scheme (RIGS) number RIGS 16-357-0521

IOP Conf. Series: Journal of Physics: Conf. Series **1018** (2018) 012002 doi:10.1088/1742-6596/1018/1/012002

Big Data and SME financing in China

Z Tian[1], A F S Hassan[2] and N H A Razak[3]

[1] Putra Business School (PBS), Universiti Putra Malaysia, 43400, Serdang, Selangor, Malaysia
[2] Faculty of Economics and Management, Universiti Putra Malaysia, 43400, Serdang, Selangor, Malaysia
[3] Faculty of Economics and Management, Universiti Putra Malaysia, 43400, Serdang, Selangor, Malaysia

E-mail: zhongkai.phd15@grad.putrabs.edu.my

Abstract. Big Data is becoming more and more prevalent in recent years, and it attracts lots of attention from various perspectives of the world such as academia, industry, and even government. Big Data can be seen as the next-generation source of power for the economy. Today, Big Data represents a new way to approach information and help all industry and business fields. The Chinese financial market has long been dominated by state-owned banks; however, these banks provide low-efficiency help toward small- and medium-sized enterprises (SMEs) and private businesses. The development of Big Data is changing the financial market, with more and more financial products and services provided by Internet companies in China. The credit rating models and borrower identification make online financial services more efficient than conventional banks. These services also challenge the domination of state-owned banks.

1. Introduction

Big Data in recent years has grown more and more popular within academia, industry fields, and even the government, and has aroused widespread concern. Researchers view Big Data as the next-generation source of economic energy [1]. Well-known management and consulting firm McKinsey claims that Big Data, as an important factor in production, has penetrated all areas of today's industry and business functions. Big Data is associated with data files, and the size of data files in the current database system is higher than the likelihood of its capture, storage, processing, and analysis. As its name suggests, Big Data involves a large amount of data, and the data collected over time is hard to analyze and process through common tools used in unified database services. This data can be employed for analyzing market trends, manufacturing, and medical science. The categories of input data include transactions histories, e-mails, photos, camera records, activity reports, and unstructured content from blogs and social media as well as data generated by various sensors. Big Data provides a description of the rapid growth and practicality of structured and unstructured data. Big Data can serve as a meaningful tool for today's Internet business and organization, as such data can lead to more accurate analysis.

Small- and medium-sized enterprises (SMEs) remain as the key driver to trigger the competitiveness, growth, innovation, and employment of European countries particularly during the economic crises when the large corporations downsize their work force [2]. The global crisis (2007-

2009) and the European sovereign debt crisis (2011-2013) have led to significant changes in financial market, especially in SME financing sector. The world's economy is more and more relying on the energy of SMEs. Not only in Europe, but also in China, SMEs provided important contribution in the fast developing economy. However, compared with large companies, SMEs faced many obstacles in the process of development, such as finance, tax, competition, and political. Finance problems is the biggest obstacles among them. The difficulty of raising capital restricts the development of SMEs.

As a developing country, China faces the problem of the lower penetration of financial services when compared with developed countries [3]. Lending services in the country disproportionately tend toward significant economic and political interests such as state-owned enterprises [4]. According to a report in 2013 from the National Bureau of Statistics of China, SMEs provide around 80% of urban employment, contribute 50% of fiscal and tax revenue, and account for 60% of the GDP in China [5]. State holding banks in China have a huge branch network, dominating the Chinese financial market. For example, as shown in Table 1 below, as of 2015, the Agricultural Bank of China (ABC) has about 24,000 branches, Industrial and Commercial Bank of China (ICBC) has about 18,000 branches, China Construction Bank (CCB) has 13,000 branches, and Bank of China (BOC) has 11,000 branches [6]. However, most SMEs in China cannot acquire enough support from these banks.

Table 1. China state holding banks' branch networks

Bank Name	Branches
ABC	24,000
ICBC	18,000
CCB	13,000
BOC	11,000

For a long time, China has been viewed as a country with an underdeveloped financial system and imperfect institutional system [7]. However, recently, China's conventional financial system has transformed into an advanced system stage by stage. With the rapid development of information and communication technology, China's financial industry has developed rapidly, especially in financial services, using Big Data and information technology to improve the financial system. At this new stage, several large financial technology companies have appeared in China. As an e-commerce company, Alibaba is becoming one of the world's largest financial technology companies. This organization is expanding its e-commerce business globally and transforming conventional payment systems [8]. However, conventional financial service companies are cautious about the new financial technology start-up companies, which are disrupting traditional finance areas

2. Big Data

Big Data is a freely defined term that describes a large number of complex data sets and describes innovative techniques for collecting and storing volumes of data [9]. In relation to information and communication technology, Big Data represents a rapidly changing and expanding area. Big Data is mainly connected to modern, trendy technologies that generate or use large amounts of data, such as semantic technology, voice and voice processing, and networking. Therefore, it is hard to define the concept of Big Data clearly. Another reason for this ambiguity is the temporal change in rapid technological development; that is, the data that represents Big Data that was difficult to approach in the 1990s can be handled by traditional computers or mobile devices through recent technological developments [10].

Today, Big Data technology represents a modern way to deal with information, which is a purpose many organizations underestimate. Big Data can be misused among organizational management; IBM's survey shows that 18% of executives only treat Big Data as a source of large amounts of data [11].

IOP Conf. Series: Journal of Physics: Conf. Series **1018** (2018) 012002 doi:10.1088/1742-6596/1018/1/012002

In general, it can be said that there are three basic characteristics of the information handled in Big Data. First of all, Big Data is too broad to process easily and in a short period of time. Second, it involves unstructured data (for example, browsing data via text, video, or audio files) and real-time information (e.g., traffic data from thousands of cameras and satellites) [10]. Third, Big Data indicates that there is a very large amount of data, and usually dealing with and managing it introduces serious logistical challenges.

Internet Data Center (IDC) defines Big Data as the next-generation architecture and technology with a cost-effective design for high-frequency, high-capacity, and diverse types of data structure. The term Big Data is used to describe and define the new era of information explosion, resulting in a lot of data, and is associated with technology development and innovation. Through the effective integration and accurate analysis of multi-source heterogeneous data, it is possible to analyze trends and better predict future event.

IBM's Big Data platform introduces the possibility of providing a large amount of data related to the organization's business adventure. This platform contains traditional technologies for handling structured and unstructured data, which links to new technologies, focusing on speed, flexibility, targeted exploration, discovery, and analysis of data. IBM defined Big Data with the following characteristics, as illustrated in Figure 1 below: Firstly, it is diverse, which means that the data is both structured and unstructured. For instance, data is taken regarding the types of messages, images, GPS signals, and other forms of data generated by the Internet and telecommunications. Second, it is gathered at great speed, which means that the data must be very quickly acquired and processed, that is, in actual or almost real-time, which allows the organization to flexibly respond to market changes or acquire a competitive advantage. Third, it demonstrates veracity, introducing the possibility of obtaining distortion output, because different data processes cause a large amount of data containing sand or distortion (such as data from a social network) simultaneously. The volume represents a large amount of data collected for analysis, which is a requirement of summary process for the organization to organize large amounts of data in a single database structure [12] [13].

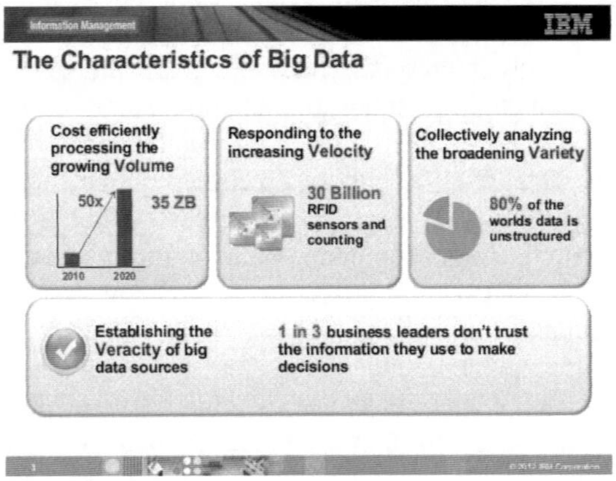

Figure.1. The Characteristics of Big Data
Source: (IBM, 2012)

Big Data comes with great value, from an information industry point of view. Big Data is the next generation of the IT industry, with a powerful driving force through cloud computing, mobile Internet, and social networks. IDC predicts that, by 2020, the third IT platform will reach $5.3 trillion in market size; from 2013 to 2020, 90% growth in the IT industry will be driven by the third IT platform [14]. Big Data will turn into a new area of economic growth, and Big Data organizations will update their eco-environments that have progressed to analytics-as-a-benefit (AaaS) models to change the IT business. Organizations, for example, IBM, Google, Microsoft, and Oracle have prepared development project on Big Data age [14].

3. SMEs Financing Problem and Big Data in China

China's economy has seen rapid development in the past three decades through the central planned economy to the market economy as the mainstay of the economy. From 1978 to 1993, China's annual GDP growth rate was 9.3 percent, 9.0 percent from 1993 to 2004, and around 9.3 percent from 2005 to 2016 (Word Bank, 2017). According to the National Bureau of Statistics of China, small- and medium-sized enterprises accounted for 99.4% of Chinese enterprises in 2012, accounting for 59% of China's gross domestic product and for 60% of total sales [15]. China's huge economic growth mainly came from the private sector in the form of small- and medium-sized enterprises, which is a major achievement of reform. However, compared with industrialized countries, China and other developing countries have a lower penetration of financial services [3]. Small- and medium-sized enterprises in emerging economies are more serious and struggle with competition from large enterprises. General consumers and commercial finance companies and microfinance institutions have achieved limited success in terms of cost-effective and sustainable demand for these groups [5].

The problems associated with SME financing come mainly from asymmetric information between lenders and SMEs. Potential default risk and lack of collateral motivate banks to refuse to lend to SMEs. There are diverse situations facing SMEs in different industries and the bank's traditional credit verification system is not compatible for all SMEs. Compared with the amount of SMEs borrowing, banks try to measure the credit cost of SMEs with the higher cost of information. High capital costs also lead banks and financial institutions to be reluctant to issue loans to SMEs.

Improving financial accessibility and affordability is thus an urgent task and issue of great concern related to economic recovery. The latest developments in computing and telecom technologies have changed the financial landscape by changing the method under which the financial industry operates to reflect an economically active microenterprise perspective [6]. Experts say that the SMEs' financial problem could be greatly remitted by creating better risk controls by using powerful computing technology and information sources [16]. A key factor in this transformation is the use of Big Data in assessing, identifying, and improving the reputation of potential borrowers and reducing transaction costs. Some possible sources could be data from social media, mobile phone usage patterns, and utility billing history [16].

Today, the world is increasingly interconnected, and a great deal of information is generated every day from social networks, search engines, and e-mail clients. Users' log files are also sent as machine-generated data like real-time monitoring for dams or bridges from sensor networks and aircrafts, cars, boats, and other vehicles. Big Data requires superior technology to efficiently handle large amounts of data efficiently. Applications for Big Data include large-scale parallel processing of databases, data mining grids, distributed file systems, distributed databases, cloud computing platforms, and scalable storage systems. Real-time or near-real-time information transfer is one of the defining features of Big Data analysis, therefore, delay of information must be avoided. Various techniques have been developed for aggregating, manipulating, analyzing, and visualizing Big Data. These technologies come from several different areas such as statistics, computer science, applied mathematics, and economics. This means that an organization that intends to obtain value from Big Data must adopt a flexible, multidisciplinary approach.

Alibaba, one of China's largest Internet companies, uses Big Data to improve risk management and control in order to improve the financing market. Alibaba developed its own credit rating and risk control model based on its payment and e-commerce transaction information. This risk control model mainly draws upon its huge online ecosystem. As of early 2015, the Alibaba Group market share consisted of more than 300 million registered users and 37 million small businesses, including Taobao and Tmall.com [17].

With the emergence of its business needs, Alibaba used optimized systems and platforms to develop advanced methods and methods to deal with its daily 10 billion users. Data processing and analysis also improved to near-real-time mode.

Alibaba, through the adjustment of Big Data technology, highlights the progress of fraud risk management. The company invented a risk management system called the Counter-Terrorist Unit

(CTU) serving as real-time payment fraud prevention monitoring system. CTU is one of the most advanced online payment fraud management systems in China. This system can track and analyze the behavior of users or accounts, identify suspicious activity, and apply different levels of treatment based on intelligent arbitration [18]. Alibaba's fraud risk management is completely different from the traditional financial and banking system. In order to deal with real-time fraud, the company has gradually developed new engineering methods to handle such data. For the hardware system, a risk prevention framework has been established to support new methodologies and algorithms. There are several different risk prevention frameworks. Alibaba uses a basic framework for fraud risk as a multi-level risk prevention framework, and Big Data make it possible to adopt real-time risk prevention online. Figure 2 illustrates the CTU operating process.

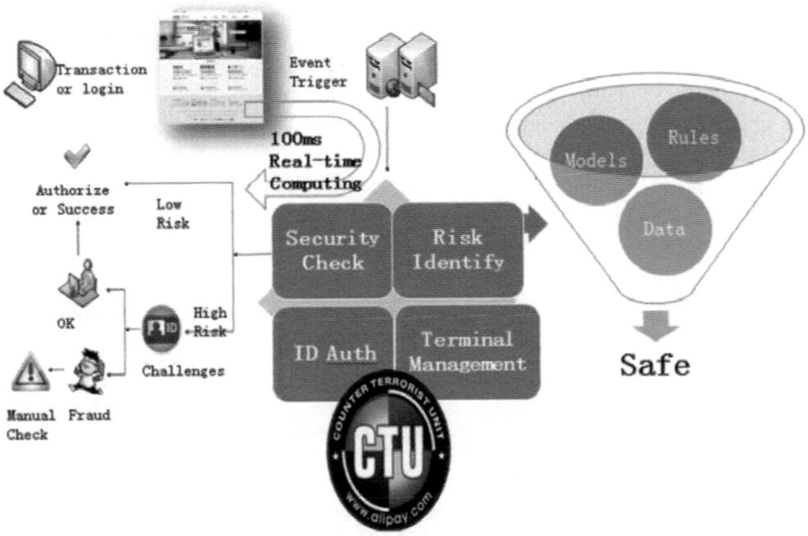

Figure.2. The Counter-Terrorist Unit (CTU)
Source: (Chen, Tao, Wang and Chen, 2015)

4. Big Data based lending services in China

Big Data has facilitated the potential transformation of China's banking sector. China Internet companies have entered the financial market and have launched various categories of financial products and services that focus on the use of Big Data. In 2015, China Rapid Finance, China's largest online consumer loan market, announced that 50 million consumers pre-approved a loan of 500 yuan (about 80 US dollars), including QQ pre-screening users, the online communication software developed by Tencent. These quotes are based on analysis data from social and financial sources to make a prediction regarding borrowers' responses and whether default or fraud is present. China Rapid Finance estimates that about 500 million consumers could be viewed as potential borrowers. The company aims to use a mobile-based channel to automatically gain credibility based on the data from several different sources [19].

Compared to traditional bank loans, the transaction costs of online financial services are lower. The Internet has greatly reduced the transaction costs associated with financial and banking business. For example, the average cost of bank transactions is estimated at $1.27, ATM transactions are $0.27, and the Internet transaction is $0.01. A study shows that no-branch banks face 38% lower expenses than commercial banks and 54% less than informal remittance channels [20]. Internet companies that provide these financial services help change the situation in China's financial markets, reducing the barriers to small businesses into the financial markets.

As of December 2013, the Alibaba cloud accumulated more than 650,000 customers and has borrowed more than 1,600 billion yuan; the average loan balance is less than 40,000 yuan, the average loan amount is 130,000 yuan, and the non-performing loan interest rate control in 1% [21]. Internet financial models, compared with traditional financial institutions, offer additional information on trade, services, costs, and other advantages. Compared with the traditional financial institutions, Internet finance models fundamentally solve the problems facing small- and medium-sized enterprises, for which financing costs are difficult problems.

MYbank, launched in 2015, for example, is an entire Internet bank and runs within the Internet environment. MYbank was established by a financial subsidiary of Alibaba. MYbank views rural customers as an important group. The company provides funds to help farmers purchase agricultural machines and tools [22]. The loans are offered up to the US $800,000 provided for SMEs and retail consumers. MYbank plans to increase its lending services to $10 million SMEs in the next five years (Bloomberg.com, 2015). The current situation has changed the traditional credit score and is more dependent on Big Data information. MYbank utilizes the service of Alibaba who launched credit evaluation companies Ant Financial and Zhima Credit, a personal credit scoring service company. These companies provide data for MYbank to evaluate borrowers. The credit information comes from Big Data such as online transactions, rented car return conditions, and if there are any court reports about default debts.

Another China Internet company, Tencent, also launched an Internet bank, WeBank, in 2015. Similar to MYbank, WeBank offers all services on the Internet with no brick-and-mortar outlets. Big Data is also involved in the operation of WeBank; borrowers just need to upload pictures from their cellphone cameras. WeBank uses Big Data to match the information of borrowers and identify their credit status. The credit information based on consumer data comes from various sources, such as online shopping behavior, activities on social network websites, online game activity, and other online behavior. The money that borrowers could obtain is based on the result of banks' analysis of credit information [23].

5. Discussion

The efforts made by companies using Big Data only capture a very small share of China's lending market, whose value was estimated at US $4 trillion in 2015 by Bloomberg.com. Nonetheless, Big Data has had a strong impact on China's financial markets. It is an encouraging sign that China's banks' interest in acquiring customer data and the lending decisions are based on economic fundamentals rather than on political considerations [6]. However, they experienced varying degrees of success and failure. It is reasonable to expect that the Chinese financial market will be reshaped by the development of Big Data.

Our critical review highlights the rapid development of Big Data in China and its application in the financial market. However, pertinent challenges relating to the role of Big Data in China financial market also remain to be explored. Such challenges could include the extent of accurate analysis by Big Data and whether Big Data can be applied to conventional financial institutions. Meanwhile, inexistence of a comprehensive rules and regulations also challenge the development of Big Data in the financial sector. Hence, future research should focus on investigating these challenges.

References

[1] Manyika J, Chui M, Brown B, Bughin J, Dobbs R, Roxburgh C and Hung A 2011 Big Data: The next frontier for innovation, competition, and productivity *Tech. rep* (McKinsey Global Institute)

[2] Moreira D 2016 The microeconomic impact on growth of SMEs when the access to finance widens: evidence from internet & high-tech industry *Social and Behavioral Sciences* 220 (2016) 278 – 287

[3] Honohan, P and King M 2009 Cause and effect of financial access: cross-country evidence from the Finscope surveys *World Bank Mimeo*

[4] Kshetri N 2011 The global financial crisis: evidence from China and India, Thunderbird *International Business Review* **53**(2), March/April pp 247–262.

IOP Conf. Series: Journal of Physics: Conf. Series **1018** (2018) 012002 doi:10.1088/1742-6596/1018/1/012002

[5] Sham T and Pang I 2014 *Monthly economic newsletter*
[6] Kshetri N 2016 Big Data's role in expanding access to financial services in China International Journal of Information Management 36 (2016) 297–308
[7] Shim Y and Shin D 2015 Analyzing China's Fintech Industry from the Perspective of Actor–Network Theory *Telecommunications Policy*
[8] Kumar A 2014 A brief history of China's Alibaba – the company that Flipkart wants to be *Your Story*
[9] Kim J 2014 Big Data Sharing Among Academics *Big Data Management, Technologies, and Applications* ISBN: 9781466647008, 177-194
[10] Cerny M 2013 *Eight technology trends that transform libraries into information society* (Osm technologickych trendu, ktere promeniknihovny v informacni spolecnosti) (in Czech)
[11] Levarsky S 2013 Seven reasons why BigData make life easier for enterprises (Sedem dôvodov, preco BigData uľahcia firmam zivot) (inSlovak)
[12] Hrapko J 2012 *How can it be used Big Data* (Ako mozno vyuziť Big Data) (in Slovak).
[13] Stanimirovic Z and Miskovic S 2014 Efficient Mataheuristic Approaches for Eploration of Online Social Networks *Big Data Management, Technologies, and Applications* ISBN: 9781466647008 p 222-269
[14] Jin X, Wah B W, Cheng X and Wang Y 2015 Significance and Challenges of Big Data Research *Big Data Research*
[15] Wang Y 2016 *Borsa Istanbul review* **16**-3 167e176
[16] Baer, Tobias, Goland T, and Schiff R 2013 *New credit-risk models for the unbanked*
[17] Alibabagroup.com 2015 Ant financial unveils China's first credit-scoring system using online data
[18] Chen J, Tao Y, Wang H and Chen T 2015 Big data based fraud risk management at Alibaba *The Journal of Finance and Data Science 1* (2015) 1-10
[19] Shu C 2015 Lending platform China rapid finance raises $35M series C at a $1B valuation
[20] McKay C and Pickens M 2010 Branchless banking pricing analysis, consultative group to assist the poor (CGAP)
[21] Wang L W and Ning X Y 2014 *Financial view*
[22] Alois 2015 J.D. MYBank, a Privately Owned Internet Bank in China, is Set to Challenge Traditional Banks
[23] Zhou X Y 2015 A small bank makes a big difference

IOP Conf. Series: Journal of Physics: Conf. Series **1018** (2018) 012003 doi:10.1088/1742-6596/1018/1/012003

The Functional Domain Analysis of TLR9 with the Visualization of the Energetically Allowed Region

Sim-Hui Tee

Multimedia University, Malaysia

shtee@mmu.edu.my

Abstract. Toll-Like Receptor 9 (TLR9) is one of the members of the toll-like receptor family that plays a pivotal role in sensing pathogen-associated molecular patterns that are harmful to human body. Upon activation, these molecular patterns trigger the downstream signaling pathways which eventually leading to immune defense. This study aims at using big data approach in exploring the functional domains of TLR9. The acquired results and analyses provide an insight for future studies in the domain of immunosurveillance.

1. Introduction

Toll-like receptors (TLRs) are a class of protein that features in the innate recognition of a wide variety of pathogens and danger signals, by detecting a specified range of pathogen-associated or damage-associated molecular patterns [1-3]. Cell-surface TLRs such as TLR4 and TLR1/2 are capable of detecting extracellular microbes; whereas intracellular TLRs such as TLR3 and TLR9 are capable of recognizing microbial nucleic acids [4]. TLRs are the most characterized pattern recognition receptors that consist of three types of domains, which are the extracellular ectodomains containing leucine-rich repeats, a trans-membrane domain, and an intracellular toll-interleukin-1 receptor domain [5]. Aberrant expression of TLRs and the malfunctioning of TLR signaling pathways are implicated in various diseases, such as autoimmune diseases [6], infectious diseases [7], inflammatory diseases [8], tumorigenesis [9], and motor abnormalities [10].

TLR9 as a member of the TLRs family is a receptor for unmethylated CpG DNA [11]. It is expressed by various immune cells such as $CD4^+$ T cells [12] and dendritic cells [13]. TLR9 ligands are single-stranded DNAs (ssDNAs) characterized with unmethylated CpG motif, which is prevalent in microorganisms [11]. Research has shown that TLR7/TLR9 signaling pathways can induce the production of type I interferons and proinflammatory cytokines [13]. The expression of TLR7 and TLR9 in B cells plays an important role in the regulation of an array of function in B cells. It was found that the balance between TLR7 and TLR9 molecules in B cells is critical in the development of B cell autoreactivity, which has profound implications in the immunity [14]. In the context of TLR7/TLR9 and B cell receptor signaling crosstalk, the degree of TLR9 responsiveness is dependent on the alterations of certain types of B-cell signaling molecules and the cellular milieu [14].

In response to viral ssDNAs in plasmacytoid dendritic cells (pDCs), TLR9 recruits MyD88, which is an adaptor protein that forms an aggregate with Interleukin-1 receptor-associated kinase 1 (IRAK-1), Interleukin-1 receptor-associated kinase 4 (IRAK-4), and Interferon regulatory factor 7 (IRF-7), all of which playing the role in transcriptional activation [15]. Phosphorylated IRF-7 translocates into the nucleus to activate the promoters of type I interferon, a type of signaling protein that is implicated in the innate immunologic defense against viral infections [16].

Despite the progress in experimental studies, large amount of data produced in the laboratory requires various computational techniques and strategies for analysis and prediction. Various computational approaches have been adopted to analyze the enormous amount of molecular data [17]. In the recent years, computational models have been built to determine the interaction of the TLR9 with CpG oligodeoxynucleotides [18]. Interaction between bacteria and toll-like receptor mediated inflammation has been modeled mathematically in the necrotizing enterocolitis context [19]. The model predicts that TLR-9 can inhibit both the beneficial and harmful effects of TLR4, a finding suggests that a synergistic balance of action between TLR4 and TLR9 is indispensable in maintaining intestinal homeostasis [19]. In this study, we strive to enhance the functional domain analysis of TLR9 using a range of computational approaches such as web services and big data analysis tools.

2. Methods

We used Protein Data Bank [20] to identify protein structural information of TLR9. A neural-network-based Pcons public web server [21] was used to predict the best matched template for TLR9. Ramachandran plot [22] and the relevant techniques in the domain of big data was used to display Ψ (backbone dihedral angles) against φ (amino acid residual sequences). We contend that the backbone of a protein to be interpreted as a class of discrete curves that permits the Frenet frames to be calculated [23]. Let

$$S_j = |P_{j+1} - P_j| \qquad (1)$$

a unit tangent vector at point P_j, where $j=0,\dots,n$-1, is given as such:

$$t_j = \frac{P_{j+1} - p_j}{S_j} \qquad (2)$$

$$P_k - P_0 = \sum_{j=0}^{k-1} s_j t_j \qquad (3)$$

3. Results and Discussion

We used a mice crystal structure of TLR9 in the complex with the inhibitory DNA4084 (form1) as our PDB input file (PDB ID: 3WPG), as obtained from NGL viewer [24]. Figure 1 demonstrates a 3D view of 3WPG.

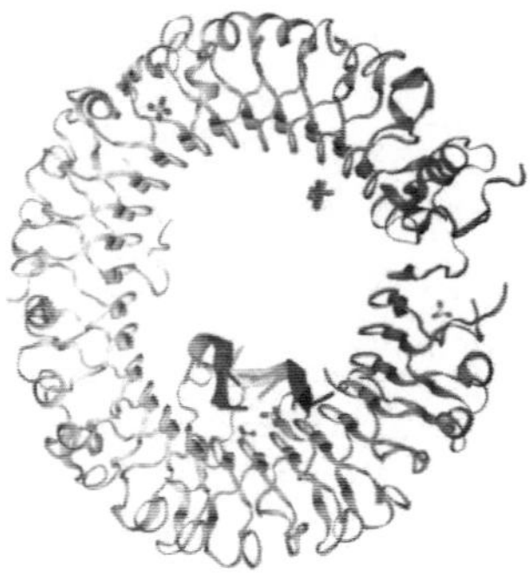

Figure 1: 3D view of 3WPG

We used the Ramachandran plot to probe the energetically permitted regions of the residues of 3WPG. We used the server hosted in the Cambridge University to perform the analysis. The plot in the figure shows each amino acid residue as a dot area in a diagram of φ against Ψ. The amino acid residues in a favored region and generously allowed regions are displayed in small red and yellow dot, respectively, in Figure 2 and Figure 3 below. We used the structure validation method developed by Lovell et al. [25].

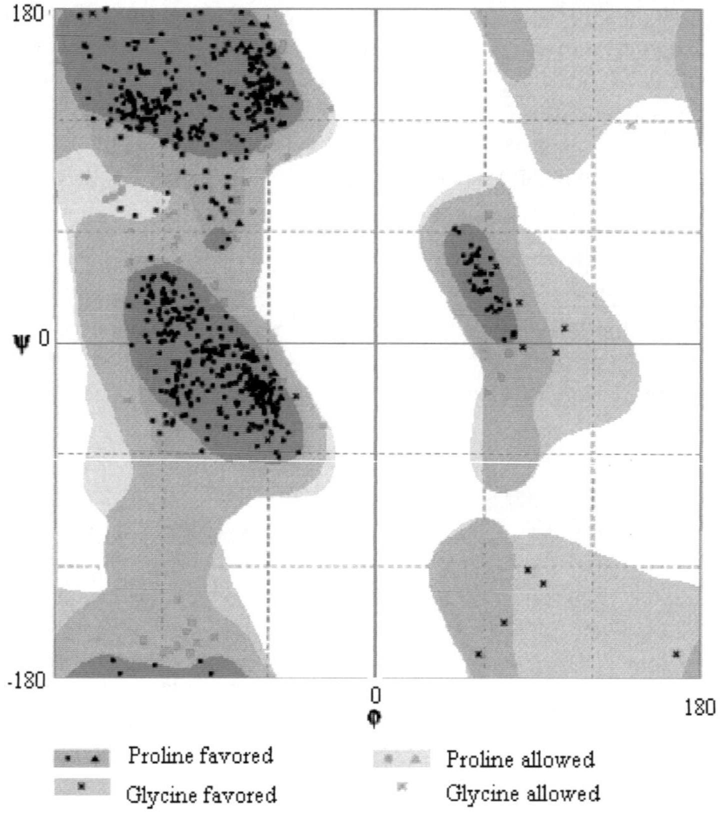

Figure 2: The 1st Ramachanran Plot for 3WPG

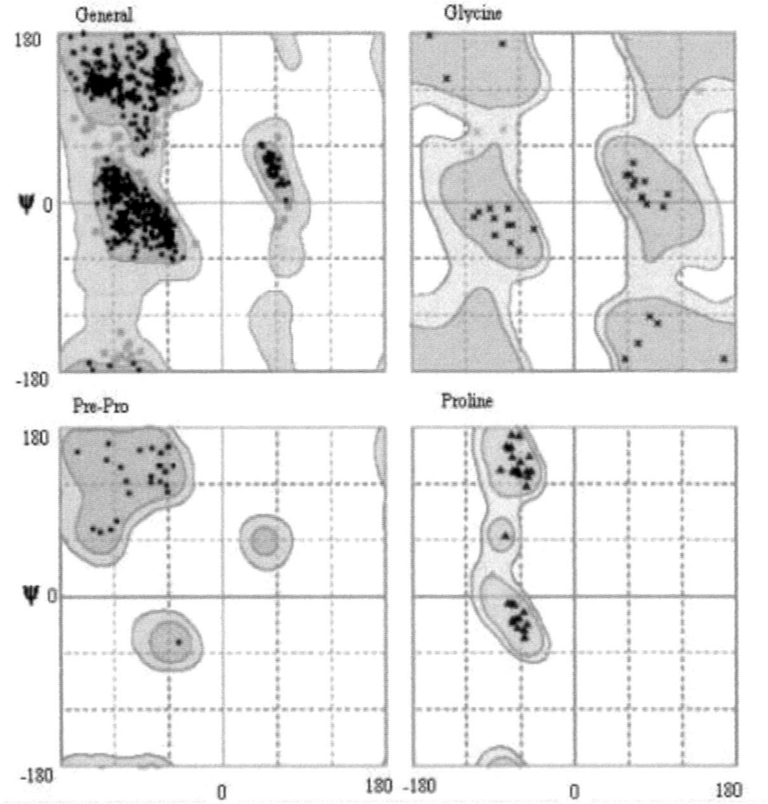

Figure 3: The 2nd Ramachanran Plot for 3WPG

In our analysis of the residues of 3WPG, 50 residues (7%) are in the allowed region. Besides, it was found that 664 residues (93%) are in the favored region of the plot. None was found in the outlier region. Due to the limit of space, only partial analyzed results are shown in Figure 4.

```
Residue [A  34 :PRO] ( -91.29,  23.82) in Allowed region
Residue [A  73 :ASN] ( -90.20,-158.41) in Allowed region
Residue [A  86 :LEU] (-110.94,  45.57) in Allowed region
Residue [A 134 :GLY] ( -77.77,  77.25) in Allowed region
Residue [A 153 :THR] (-115.54,-165.53) in Allowed region
Residue [A 174 :MET] (-147.90,  78.79) in Allowed region
Residue [A 178 :CYS] (-151.95,  77.42) in Allowed region
Residue [A 206 :LEU] (-138.06,  60.52) in Allowed region
Residue [A 230 :ASN] (-111.62,-138.32) in Allowed region
Residue [A 242 :GLN] (  73.36,  -5.27) in Allowed region
Residue [A 255 :CYS] (-142.55,  82.97) in Allowed region
Residue [A 286 :LEU] ( -59.77, 107.68) in Allowed region
Residue [A 308 :VAL] ( -84.84,  22.32) in Allowed region
Residue [A 310 :LEU] ( -59.23, 110.02) in Allowed region
Residue [A 318 :ASN] (-116.36,-165.42) in Allowed region
Residue [A 319 :PHE] (-112.69,  70.79) in Allowed region
Residue [A 327 :THR] ( -51.60, 160.59) in Allowed region
Residue [A 333 :LEU] ( -69.14,  71.17) in Allowed region
Residue [A 375 :PHE] (  61.96,  69.29) in Allowed region
Residue [A 383 :THR] ( -29.32, -44.18) in Allowed region
Residue [A 393 :LEU] ( -64.14, 104.81) in Allowed region
Residue [A 425 :ASN] (-130.38,-160.13) in Allowed region
Residue [A 469 :ASP] ( -72.70,  86.23) in Allowed region
Residue [A 470 :ARG] (-163.07,  89.58) in Allowed region
Residue [A 471 :CYS] (  68.56, -18.54) in Allowed region
Residue [A 473 :ASN] (  62.10, -26.26) in Allowed region
```

Figure 4: Partial results of the analysis for the residues

4. Conclusion

TLR9 is a critical pathogen pattern detection protein in human body. A complete insight of its structural, biochemical, and molecular features is vital in molecular medicine and applied research. In this paper, the functional domains of TLR9 was probed using computational tools coupled with the big data approach. Residue properties were analyzed to validate the structure of 3WPG, which is a crystal structure of TLR9 in mouse models. Ramachandran plots show that most of the amino acid residues in our study emerge in the most favored regions. The molecular characteristics of 3WPG thus provide insights into the functional domains of TLR9.

References

[1] S. Akashi-Takamura and K. Miyake, "TLR accessory molecules," Current Opinion in Immunology, vol. 20, pp. 420-425, 2008.

[2] G. Trinchieri and A. Sher, "Cooperation of toll-like receptor signals in innate immune defence," Nature Reviews Immunology, vol. 7, pp. 179-190, 2007.

[3] T. Kawai and S. Akira, "The role of pattern-recognition receptors in innate immunity: update on toll-like receptors," Nature Immunology, vol. 11, pp. 373-384, 2010.

[4] T. Kaisho and S. Akira, "Toll-like receptor function and signaling," Journal of Allergy and Clinical Immunology, vol. 117, pp. 979-989, 2006.

[5] T. Kondo, T. Kawai, and S. Akira, "Dissecting negative regulation of toll-like receptor signaling," Trends in Immunology, vol. 33, pp. 449-458, 2012.

[6] M. Ehlers and J. Ravetch, "Opposing effects of Toll-like receptor stimulation induce autoimmunity or tolerance," Trends in Immunology, vol. 28, pp. 74-79, 2007.

[7] V. Balloy, M. Si-Tahar, O. Takeuchi, B. Philippe, M-A. Nahori, M. Tanguy, M. Huerre, S. Akira, J-P. Latgé, and M. Chignard, "Involvement of Toll-like receptor 2 in experimental invasive pulmonary aspergillosis," Infection and Immunity, vol.73, pp. 5420-5425, 2005.

[8] K.H. Shalaby, S. Al Heialy, K. Tsuchiya, S. Farahnak, T.McGovern, P.A. Risse, W.K. Suh, S.T. Qureshi, and J.G. Martin, "The TLR4-TRIF pathway can protect against the development of experimental allergic asthma," Immunology, 2017, doi:10.1111/imm.12755

IOP Conf. Series: Journal of Physics: Conf. Series **1018** (2018) 012003 doi:10.1088/1742-6596/1018/1/012003

[9] I. Hirsch, C. Caux, U. Hasan. N. Bendriss-Vermare, and D. Olive, "Impaired Toll-like receptor 7 and 9 signaling: from chronic viral infections to cancer," Trends in Immunology, vol.31, pp. 391-397, 2010.

[10] V. Patel, A.M. Patel, and J.J. McArdle, "Synaptic abnormalities of mice lacking toll-like receptor (TLR)-9," Neuroscience, vol. 324, pp. 1-10, 2016,

[11] H. Hemmi, O. Takeuchi, T. Kawai, T. Kaisho, S. Sato, H. Sanjo, M. Matsumoto, K. Hoshino, H. Wagner, K. Takeda, and S. Akira, "A Toll-like receptor recognizes bacterial DNA," Nature, vol. 408, pp. 740-745, 2000.

[12] J. Reynolds and C. Dong, "Toll-like receptor regulation of effector T lymphocyte function," Trends in Immunology, vol. 34, pp. 511-519, 2013.

[13] K. Hoshino and T. Kaisho, "Nucleic acid sensing Toll-like receptors in dendritic cells," Current Opinion in Immunology, vol. 20, pp. 408-413, 2008.

[14] A.N. Suthers and S. Sarantopoulos, "TLR7/TLR9- and B cell receptor-signaling crosstalk: Promotion of potentially dangerious B cells," Frontiers in Immunology, 8:775, doi:10.3389/fimmu.2017.00775, 2017.

[15] T. Kawai, S. Sato, K.J. Ishii, C. Coban, H. Hemmi, M. Yamamoto, K. Terai, M. Matsuda, J. Inoue, S. Uematsu, O. Takeuchi, and S. Akira, "Interferon-alpha induction through Toll-like receptors involves a direct interaction of IRF7 with MyD88 and TRAF6," vol. 5(10), pp. 1061-1068, 2004.

[16] G.A. Versteeg and A. García-Sastre, "viral tricks to grid-lock the type I interferon system," Current Opinion in Microbilogy, vol. 13, pp. 508-516, 2010.

[17] J. Konc and D. Janežič, "ProBiS-2012: web server and web services for detection of structurally similar binding sites in proteins," Nucleic Acids Research, vol. 40, pp. W214-W221, 2012.

[18] W. Zhou, Y. Li, X. Pan, Y. Gao, B. Li, Z. Qiu, L. Liang, H. Zhou, and J. Yue, "Toll-like receptor 9 interaction with CpG ODN: An in silico analysis approach," Theoretical Biology and Medical Modelling,10:18, 2013.

[19] J. Arciero, E.G. Bard, R. Siggers, A. Afrazi, D. Hackam, Y. Vodovotz, and J. Rubin, "Modeling the interactions of bacteria and Toll-like receptor-mediated inflammation in necrotizing enterocolitis," Journal of Theoretical Biology, vol. 321, pp. 83-99, 2013.

[20] H. Berman, J. Westbrook, Z. Feng, G. Gilliland, T. Bhat, H. Weissig, I. Shindyalov, and P. Bourne, "The Protein Data Bank," Nucleic Acids Research, vol 28, pp. 235-242, 2000.

[21] J. Lundstrom, L. Rychlewski, J. Bujnicki, and A. Elofsson, "Pcons: a neural-network-based consensus predictor that improves fold recognition," Protein Science, vol. 10, pp. 2354-2362, 2001.

[22] G. Ramachandran, C. Ramakrishnan, and V. Sasisekharan, "Stereochemistry of polypeptide chain configurations," Journal of Molecular Biology, vol. 7, pp. 95-99, 1963.

[23] J. Quine, T. Cross, M. Chapman, and R. Bertram, "Mathematical aspects of protein structure determination with NMR orientational restraints," Bulletin of Mathematical Biology, vol. 66, pp. 1705-1730, 2004.

[24] A.S. Rose, A.R. Bradley, Y. Valasatava, J.M. Duarte, A. Prlić, and P.W. Rose, "Web-based molecular graphics for large complexes," ACM Proceedings of the 21st International Conference on Web 3D Technology (Web3D '16), pp. 185-186, 2016.

[25] S.C. Lovell, I.W. Davis, W. Arendall, P. de Bakker, J. Word, M.G. Prisant, J. Richardson, D.C. Richardson, "Structure validation by Calpha geometry: phi, psi and Cbeta deviation," Proteins, vol 50(3), pp. 437-450, 2003.

IOP Conf. Series: Journal of Physics: Conf. Series **1018** (2018) 012004 doi:10.1088/1742-6596/1018/1/012004

A Comprehensive Review of Existing Risk Assessment Models in Cloud Computing

Ahmad Amini, Norziana Jamil

Institute of Informatics and Computing in Energy (IICE), Universiti Tenaga Nasional, Jalan IKRAM-UNITEN,Selangor, Malaysia.
Email: Ahmad.Amini2015@gmail.com, Norziana@uniten.edu.my

Abstract. Cloud computing is a popular paradigm in information technology and computing as it offers numerous advantages in terms of economical saving and minimal management effort. Although elasticity and flexibility brings tremendous benefits, it still raises many information security issues due to its unique characteristic that allows ubiquitous computing. Therefore, the vulnerabilities and threats in cloud computing have to be identified and proper risk assessment mechanism has to be in place for better cloud computing management. Various quantitative and qualitative risk assessment models have been proposed but up to our knowledge, none of them is suitable for cloud computing environment. This paper, we compare and analyse the strengths and weaknesses of existing risk assessment models. We then propose a new risk assessment model that sufficiently address all the characteristics of cloud computing, which was not appeared in the existing models.

1. Introduction

During these few years the use of cloud computing by organizations and individuals have increased as they gain access to a shared resources. Cloud computing enables the adoption of on-demand services with virtualized resources through minimal management effort and service provider's interaction [1] . The flexibility and elasticity of this new paradigm can be seen when different cloud services can be integrated using computing resources that benefits users especially when they face restricted economy resources. On the other hand, cloud computing by outsourcing resources that beyond organization boundaries, enforces users to host strategic assets in an inaccessible area. This becomes the main source of risk, which needs to be mitigated in order to guarantee the service's assurance [2]. The lack of control and trust, and the concern of data confidentiality and data integrity, has forced organizations to perform risk assessment to determine their privacy and security. This is done by assessing threats and analysing vulnerabilities. Risk assessment is an integral part of risk management and it is aimed to assess the threat's impacts [3]. The circumstance of assessing risk depends on different factors which are cloud's services, cloud's models, and the required information security (accessibility, availability, integrity, confidentiality).

These challenges lead many researches to develop quantitative or qualitative based assessment models to produce evaluated results to be used by organization as a guide to secure and protect their outsourced assets. However, cloud computing still lacks of standardized information security framework, which applies to risk assessment as well [1]. This paper focuses on a specific aspect of risk assessment such as threat, vulnerability, assets identification and cloud computing characteristics such as rapid elasticity, on-demand self-services, resource pooling, broad network access, and measured service. The result of this research leads to the development of a quantitative risk assessment model to aid organizations in establishing an accurate security risk plan to protect critical assets and gather requirements before applying risk assessment. This paper is organized as follows: Section 2 overview risk assessment in general to give an idea the fundamental steps involved in risk assessment. The understanding of risk assessment framework is important so that it helps to minimize the risk of threats. We detailed out the methodology that we used to conduct this study in Section 3 and our proposal of risk assessment framework is given in Section 4.

2. Risk Assessment

In terms of information security, due diligence should be undertaken to ensure risks are managed properly that includes a process of identifying and assessing security risks [4]. The whole aim is to protect organization's assets. This process is known as security risks management that involves identifying critical assets and threats, and assessing vulnerability and security risk [5] . Risk management framework in cloud computing consists of five stages: user requirement self-assessment, cloud service provider desktop assessment, risk assessment, third party agencies review, and continuous monitoring [6]. This study is only focusing on the third stage of risk management framework, which is risk assessment. This stage is considered to be the most crucial stage in risk management. We reviewed existing models of risk assessment to determine potential threats and risks that are associated with cloud computing. On the other word, on-demand nature of cloud computing is the source of critical threat's impacts such as data lose or leakage, account or service hijacking, denial of service, and data breaches [7]. Risk assessment aims to analyse potential risk of information systems and supply effective strategies for risk reduction through recognizing assets, threats and vulnerabilities [8].

Unfortunately, most proposed risk security assessment models do not guarantee obtaining risks in novel and intangible environments. Cloud computing characteristics are the same as on-demand self-service, broad network access, resource pooling, rapid elasticity and measured service which make it attractive and also tend to make it hard to assess its security risks [9] . Accordingly, through cloud computing evaluation, new risks are different from one organization to another. For this reason, risk assessment strategies have been developed and integrated into cloud organization based on business aspects. In this manner, the level of risk in many cases, vary significantly with the type of cloud environment being considered [10]. Nowadays, several numbers of risk assessment, framework, and methodology are available to eliminate the possible risks and potential threats that are associated with an IT systems [11]. The risk identification is the first step of risk assessment process. In cloud computing the main source of risks are malicious threats to exploit vulnerabilities. These threats are divided into general threats that are applicable to any type of information system and unique threats that can affect cloud's users and providers. These threats are associated with virtualization technology, data protection, controlling and allocating resources. The next step is analysing risk to determine the impact and likelihood of each risk. Measuring and assigning value to each risk is the last step of risk assessment approach that is done through quantitative, qualitative or integration of both methods. However, the list of risks and their value can be different base on services, models, or other characteristics of cloud computing.

3. Research Methodology

The methodology of this research is divided into four phases namely Phase 1: review cloud computing characteristics and risks, Phase 2: review existing risk assessment models in cloud computing, Phase 3: compare advantage and disadvantages of existing risk assessment models, and Phase 4: define risk assessment approaches. In Phase 1, we review cloud computing characteristics and different risks that organizations encounter when adopting cloud computing. Different risk assessment models are then reviewed in Phase 2. All models were chosen based on their possibility to be adopted into cloud computing. The details of every phase is given below.

3.1. Phase 1: Define Cloud Computing (CC) characteristics

For assessing the security risks that could affect cloud computing environment, we have a closer look at cloud computing characteristics which affect the risk assessment process.

- ***On-demand services:*** A user can request and manage one or more services whenever he/she needs and pay as "pay-and-go" without any human interaction with a Cloud Service Provider (CSP) [12].
- ***Rapid elasticity:*** Resources can elastically and rapidly provisioned as customer's demands and they can have unlimited resources to be purchased based on their requests [13].
- ***Broad access network:*** The availability of resources and customer's data that are located in different virtual machines creates a dynamic collection of resources that can be harmed by unauthorized users.
- ***Resource pooling:*** Computing resources are combined to serve multi customers by different physical and virtual resources which are dynamically assigned and reassigned based on demands [14] .
- ***Metered services:*** Measuring and reporting the usage of system dynamically, is a major abstraction in cloud computing in pay-as-you-use manner.

The above five properties of cloud computing, highlight the dynamic nature of cloud computing that requires a dynamic and cloud-oriented risk assessment approach to identify the risks of cloud adopted assets.

3.2. Phase 2: Reviewing existing Risk Assessment Models

Assessing risks in cloud computing as a complex and dynamic environment has additional challenges. The lack of undefined clear boundaries between cloud services, enforce businesses to develop dynamic risk assessment instead of statistic approaches in cloud computing. Consequently, different models have been developed to modify risk assessment models and prevent users and providers from hazardous risks. The existing Risk Assessment models reviewed are briefed in the following sections.

3.2.1. OCTAVE

Operationally Critical Threat, Asset, and Vulnerability Evaluation (OCTAVE) is the methodology to identify and evaluate information security risks by developing qualitative risk evaluation criteria and identifying assets, vulnerabilities and risks. This approach consists of eight steps that are organized into four groups (Figure 1). The primary benefit of OCTAVE is connecting organizational objectives to information risk assessment and vulnerability management. For this reason, OCTAVE is developed as flexible and self-modified approach to be customized for different organizations by their small teams of IT and the business. However, this model can be integrated by other models and used as a hybrid model that it seems as a big opportunity for cloud computing. But unfortunately, documenting all requirements and characteristics of this integrated approach, makes this model more complex for distributed and massive computing environments. The method of threat classification and identification is another disadvantage of OCTAVE model.
Also, the threat tree as a guide for identifying threats in OCTAVE is not suitable for massive computing environment because the users are faced with a huge and unclear tree by many paths where each path presents a real-word scenario. This threat identification approach is a single-dimension and furthermore static in that it cannot support different criteria (agent, source, impact and motivation) of threat classification and identification in cloud computing. The elimination of vulnerability identification tool is another issue in OCTAVE Allegro. It seems this function of OCTAVE is more suitable for static environment that company can integrate their own vulnerability classification tool. Classification of vulnerabilities in cloud computing has to follow principles such as public acceptance, comprehensibility, competence, determinism, mutual exclusion, and repeatability [15] by respecting Web Service (WS), Service-Oriented-Architecture (SOA), and virtualization as the main core of cloud computing. By considering all the above mentioned criteria and functionalities of vulnerability

assessment, it seems is more appropriate and efficient to integrate vulnerability identification into cloud services.

3.2.2. Semi-quantitative BLO-oriented cloud risk assessment (SEBCRA)

This model as depicted in Figure 2 evaluates the impact of risks on an organization's global Business-Level Objectives (BLO) rather than considering the effect on different single assets of organization [10]. Semi-quantitative or hybrid is the primary method to assess risk in this model, which is less numerical than quantitative methods and risks have been classified based on their probability. The core concept is based on risk level estimation for each BLO that is proportional to the probability of a given risk and its impact on the BLO [16]. The qualitative part is risk level estimation by risk-level matrix that includes probability of risk occurrence (very unlikely, possible, likely, and frequent) and impact of each risk (very high, high, medium, low, and very low).

The outcome of this matrix is a list of prioritized risk that is obtained by multiplying probability with impact of risk to maximize profit and user satisfaction. Known and unknown risk identification is the most critical part of risk assessment method used in this model. It is so because the risk identification is based on service provider and impact of risks that is so narrow. This indicates that additional factors have to be investigated such as motivation and source of risks. On the other hand, multi dimensions model is more efficient for on-demand and massive computing environment to identify and

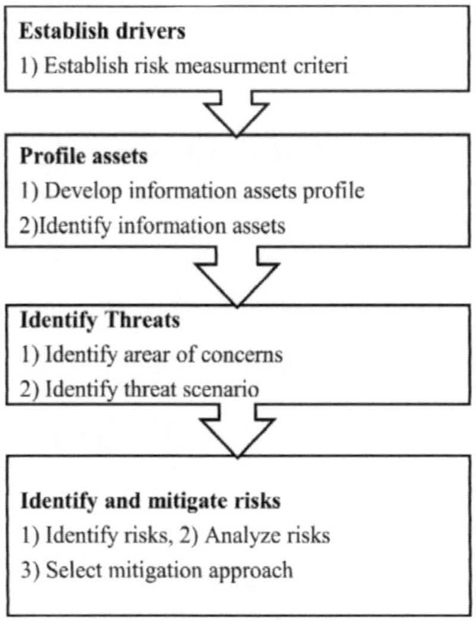

Figure 1: OCTAVE Model

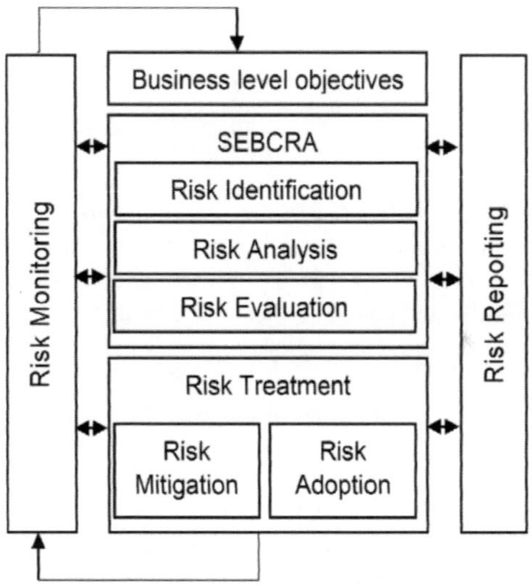

Figure 2: LO driven cloud risk management

classify risks. Measuring the impact of each risk is related to main information security principles such as integrity, confidentiality and availability. These impact metrics allow provider to calculate accurately the cost of each risk for the system. In this model, it is more important to measure the cost of threat quantitatively by considering security metrics rather than classifying them by qualitative matrix.

3.2.3 A cyber security model

This model proposes security metrics (stakes, dependency and impact matrix, security requirements, component and services, security threats) that are quantified in economic terms to enable service provider and service subscriber to quantify the risks that they incur as a result of prevailing security threats and system vulnerabilities [17]. This quantitative model, summarized all metrics in matrix of mean failure cost to measure threats. On the other hand, the discussed model allows providers and subscribers to make security decision based on quantitative analysis.

3.2.4. Scalable Risk Assessment method using game theory (CCRAM)
Game theory is an interdisciplinary approach to inspect the behaviour between two players or groups and find optimal strategies to increase outcome or decrease damage of risks [18] .The main purpose of CCRAM model is to allow defender to choose a defence strategy against attacker's action. For modelling risk assessment, Confidentiality, Integrity, and Availability (CIA) are considered to calculate the impact of attack on the system. This is assuming virtualization and live migration of resources and assets in cloud computing affect the value of CIA from one system to another. However, in CCRAM model, the quantitative values are used to calculate the risk's effects and determine which one is more important. The three main characteristics of risk assessment model are listed as impact of threat, likelihood, probability and value of assets. The total value of asset and the level of damage by risks are represented in this model but the level of system's vulnerabilities to calculate the probability or likelihood is not mentioned in the above model.

3.3. Phase 3: Comparing the advantages and disadvantages of presented risk assessment approaches
The comparison of different discussed models is summarized in Table 1. The advantages and disadvantages of different risk assessment models are based on risk assessment and cloud computing characteristics. The first step of this comparative study is to specify the main characteristics of all risk assessment methodologies. In the early stage, organizations should clarify which assets are more important and critical for them which could be classified such as private-classified, financial-classified and application assets-classified [19].

The next step is to identify threats and vulnerabilities and to consider how threats can exploit vulnerabilities. For this reason, effective security classification is necessary to identify threats and vulnerabilities into classes based on the intended effect of attacks and develop solutions to prevent system from hazardous threats [20]. In this situation, organization which are interested to adopt cloud computing have to modify and present the combination list of threats and map them to indicate vulnerabilities and clarify the security concerns that are effected by each threat. Finally, the presented list have been used to estimate the security risk based on impact and likelihood.

However, comparison of the risk assessment models presents great opportunities to look at similarities in model's infrastructure and develop a novel risk assessment model. The next step of the comparative study is about sub-characteristics of main risk assessment characteristics. Each sub-feature has been presented as the main objective of a novel risk assessment in cloud computing. Furthermore, we try to carry the comparison among elements and cross-checked and marked against all proposed models in the table 2. Nonetheless, all the compared results are very useful to quantify the security risk assessment for each organization and translate them to a homogenized risk analysis result.

3.4. Phase 4: define the risk assessment approaches
For migrating organizations' information to the cloud the optimal rules have to be investigated to apply some approaches in order to reduce cost and improve security. Most of the risk assessment models in our comparison are not suitable for assessing risk in complex and dynamic environments such as cloud computing. The most important holistic approaches to address the above challenges are discussed below:

- Organization that are interested to adopt cloud computing into their business must rely essentially on automated mechanisms. In fact, based on cloud computing's promises automated control mechanisms are supposed to be more effective. As a result, the risk assessment model has to be an automated system with less training and better risk managing and privacy monitoring.

- The integration of different dynamic computing environments by different properties and characteristics, supporting virtualization and resource pooling are challenges and concerns for assessing risks in traditional assessment methodologies. In this situation, defining boundaries and controlling the access to virtual resources can be defined as risk assessment approaches to aid organization to protect system from cyber threats when adopting it into the cloud computing.

  By sharing physical and virtual resources between different consumers in cloud computing, a traditional risk assessments have to be updated to support resource pooling and multi-tenancy. In this shared resources environment, each consumer can carry out malicious activities that may affect other consumers. Therefore, controlling the access to different resources and automated resource allocation have to be considered in order to develop and adopt risk assessment in cloud computing.

Table 1. Comparing risk assessment models based on cloud computing characteristics

RA models	**Risk assessment characteristics**		**Cloud computing charactristics**			
	Threat identification and classification	**Measured services**	**On-demand self-services**	**Broad network access**	**Resource pooling**	**Rapid elastisity**
OCTAVE	Use threat tree as a single dimension	The lack of suitable threat and vulnerability identification	customized model	lack of clearly-defined bounderies	Need more modification for multi-tenancy environment and addressing unknown location	Slightly strong to migrate by another risk assessments
SEBCBRA	use risk level matrix	costing and measuring VMs follows pay-as-you-go model	Use Business-Level Agreement	Mean Time Between Failure (MTBF) and Mean Time to repair (MTTF) to assess the availability of resources	Offer round-robin schaduler to ensure equitable distribution of VMs	Use workload pattern for real access to resources
A cyber security model	Use main failure cost matrix	Calculate the security gain cost	self-managed purpose	access to a computational resources	provide allocation mechanisms	resource migration, communication,and mobility
CCRAM	Formulate CIA value to measure the impact of	The lack of pay-as-you-use	out of self-managed process and need expert	Support live migration	lack of resource housting and multi-tenancy	Identify cloud live migration

Asset's value estimation	**Vulnerability identification and classification**
Identify and develop profile of asset's information	Lack of suitable tool to identify vulnerability
Consider the business and IT-level objective as assets	Lack of vulnerability assessment
Estimate the asset's value	Define virtualization as the source of major security risks
The damage level of assets is formulated in this model based on CIA form	Miss the evaluation of vulnerabilities

- By giving opportunity to users to scale up and down resources elasticity and rapidly, the effective risk assessment model must cover providers, sub-providers, and consumers to optimize the use of available and shared resources. For this reason, risk assessment has to be able to migrate to another risk assessment models and control the live migration of physical and virtual resources dynamically.
- Controlling and optimizing the used resource by consumers, enforce organizations to deploy risk assessment model to control the collected confidential data of users that are potentially vulnerable. In this situation, protecting and monitoring the metering information of each individual user is an important approach for adopting risk assessment into cloud computing.

Table 2. Comparative of sub characteristics among different risk assessment models

Risk assessment		**Risk assessment models**			
Characteristics	**Sub-characteristics**	**OCTAV**	**SEBCBRA**	**A cyber securitv**	**CCRAM**
1. On-demand self-services	Simple and self-service user portal	✓	✓	×	×
2.Broad network access	Clearly-defind boundry	×	✓	?	?
	Access to resources as virtual	×	✓	✓	✓
	Identity and Access Management	✓	✓	✓	✓
	Federated differend services	✓	?	?	✓
3.Resource pooling	Access of user to different resource	?	?	✓	?
	Automated allocation and re-	✓	?	?	✓
4.Rapid elastisity	Live migration	✓	✓	✓	✓
	Control Physical and virtual resource	×	?	✓	×
	Risk assessment migration	✓	×	×	×
	Resource migration and	×	✓	✓	×
5.Measured services	Used-resource calculation and	×	✓	✓	×
6. Threat identification and classification	Threat identification	✓	✓	✓	✓
	Threat impact measurement	×	✓	✓	✓
7. Vulnerability identification and	Vulnerabilities identification	✓	×	✓	×
8. Asset's value estimation	Assets identification and value	✓	✓	✓	✓

Full fill requirements: ✓, doesn't full fill : ×, couldn't find documentation: ?

4. Conclusion

The adoption of cloud computing provides continuous benefits such as access to unlimited virtualized resources, management and maintenance of system professionally, access to on-demand services and billing through as pay-as-you-need. Unfortunately, these advantages do not offer better security in terms of integrity, confidentiality, and availability. For this reason, risk assessment models have been developed to enable service providers and users to quantify risks based on their impact and severity on critical assets. Moreover, developing risk assessment for cloud computing in a complicated environment requires comprehensive quantitative and qualitative metrics. The most of the risk factors and metrics will rely on decision maker's or expert's judgments. In this work, we investigate different risk assessment models by carefully considering cloud computing characteristics and considering the main features of risk assessment model. We do a classification of risk and obtain a generic baseline to develop a new risk assessment model that suits well in cloud computing. We envision to extend this work to refine a quantitative and qualitative risk assessment approach for measuring and ranking risks for cloud computing environment.

Acknowledgement

First I would like to thank to my supervisor Dr. Norziana Jamil for her guidance and her advice. This study is partly funded by TNB Seed Fund 2016 under a project 'R&D of Micro Cryptographic Engine for Protecting Data in SCADA Link'. I would specially thank my family who has supported me through the ups and downs in completing this study. Without helps of the person mentioned above, I couldn't be succeed in this research.

References

[1] Theoharidou, M., N. Tsalis, and D. Gritzalis, *In Cloud We Trust: Risk-Assessment-as-a-Service*, in *Trust Management VII*, C. Fernández-Gago, et al., Editors. 2013, Springer Berlin Heidelberg. p. 100-110.

[2] Rebollo, O., et al., *Empirical evaluation of a cloud computing information security governance framework.* Information and Software Technology, 2015. **58**: p. 44-57.

[3] Peiyu, L.I.U. and L.I.U. Dong, *The New Risk Assessment Model for Information System in Cloud Computing Environment.* Procedia Engineering, 2011. **15**: p. 3200-3204.

[4] Humphreys, E., *Information security management standards: Compliance, governance and risk management.* Information Security Technical Report, 2008. **13**(4): p. 247-255.

[5] Bajpai, S., A. Sachdeva, and J.P. Gupta, *Security risk assessment: Applying the concepts of fuzzy logic.* Journal of Hazardous Materials, 2010. **173**(1–3): p. 258-264.

[6] Khrisna, A. and Harlili. *Risk management framework with COBIT 5 and risk management framework for cloud computing integration.* in *Advanced Informatics: Concept, Theory and* Application (ICAICTA), 2014 International Conference of. 2014.

[7] Amini, A., et al., *A Fuzzy Logic Based Risk Assessment Approach for Evaluating and Prioritizing Risks in Cloud Computing Environment*, in Recent Trends in Information and Communication Technology: Proceedings of the 2nd International Conference of Reliable Information and Communication Technology (IRICT 2017), F. Saeed, et al., Editors. 2018, Springer International Publishing: Cham. p. 650-659.

[8] Guan, J.-z., et al., *Knowledge-based information security risk assessment method.* The Journal of China Universities of Posts and Telecommunications, 2013. 20, Supplement 2: p. 60-63.

[9] Burton S. Kaliski, J. and W. Pauley, *Toward risk assessment as a service in cloud environments*, in *Proceedings of the 2nd USENIX conference on Hot topics in cloud computing*2010, USENIX Association: Boston, MA. p. 13-13.

[10] Fitó, J.O. and J. Guitart, *Business-driven management of infrastructure-level risks in Cloud providers.* Future Generation Computer Systems, 2014. **32**: p. 41-53.
[11] Saleh, M.S. and A. Alfantookh, *A new comprehensive framework for enterprise information security risk management.* Applied Computing and Informatics, 2011. **9**(2): p. 107-118.
[12] Jula, A., E. Sundararajan, and Z. Othman, *Cloud computing service composition: A systematic literature review.* Expert Systems with Applications, 2014. **41**(8): p. 3809-3824.
[13] Ali, M., S.U. Khan, and A.V. Vasilakos, *Security in cloud computing: Opportunities and challenges.* Information Sciences, 2015. **305**: p. 357-383.
[14] Sher DeCusatis, C.J. and A. Carranza, *Chapter 15 - Cloud Computing Data Center Networking*, in *Handbook of Fiber Optic Data Communication (Fourth Edition)*, C. DeCusatis, Editor. 2013, Academic Press: Oxford. p. 365-386.
[15]Rabai, L.B.A., et al., *A cybersecurity model in cloud computing environments.* Journal of King Saud University - Computer and Information Sciences, 2013. **25**(1): p. 63-75.
[16]Djemame, K., et al., *A Risk Assessment Framework and Software Toolkit for Cloud Service Ecosystems*, in *The Second International Conference on Cloud Computing, GRIDs, and Virtualization*2011, IARIA. p. 119-126.
[17] Rabai, L.B.A., et al., *A cybersecurity model in cloud computing environments.* Journal of King Saud University - Computer and Information Sciences, 2013. **25**(1): p. 63-75.
[18] Furuncu, E. and I. Sogukpinar, *Scalable risk assessment method for cloud computing using game theory (CCRAM).* Computer Standards & Interfaces, 2015. **38**: p. 44-50.
[19] Amini, A., et al., *Threat Modeling Approaches for Securing Cloud Computin.* Vol. 15. 2015. 953-967.
[20] Jouini, M., L.B.A. Rabai, and A.B. Aissa, *Classification of Security Threats in Information Systems.* Procedia Computer Science, 2014. **32**: p. 489-496.

IOP Conf. Series: Journal of Physics: Conf. Series **1018** (2018) 012005 doi:10.1088/1742-6596/1018/1/012005

Mobile Cloud Computing with SOAP and REST Web Services

Mushtaq Ali [*,1], **Mohamad Fadli Zolkipli** [2], **Jasni Mohamad Zain** [3], **Shahid Anwar** [4]

[1,2,4] Faculty of Computer Systems & Software Engineering. Universiti Malaysia Pahang Gambang, 26300, Malaysia
[3] Faculty of Computer & Mathematical Sciences, Universiti Teknology MARA (UiTM), 40450 Shah Alam, Selangor, Malaysia sunnyswa@hotmail.com

Abstract—Mobile computing in conjunction with Mobile web services drives a strong approach where the limitations of mobile devices may possibly be tackled. Mobile Web Services are based on two types of technologies; SOAP and REST, which works with the existing protocols to develop Web services. Both the approaches carry their own distinct features, yet to keep the constraint features of mobile devices in mind, the better in two is considered to be the one which minimize the computation and transmission overhead while offloading. The load transferring of mobile device to remote servers for execution called computational offloading. There are numerous approaches to implement computational offloading a viable solution for eradicating the resources constraints of mobile device, yet a dynamic method of computational offloading is always required for a smooth and simple migration of complex tasks. The intention of this work is to present a distinctive approach which may not engage the mobile resources for longer time. The concept of web services utilized in our work to delegate the computational intensive tasks for remote execution. We tested both SOAP Web services approach and REST Web Services for mobile computing. Two parameters considered in our lab experiments to test; Execution Time and Energy Consumption. The results show that RESTful Web services execution is far better than executing the same application by SOAP Web services approach, in terms of execution time and energy consumption. Conducting experiments with the developed prototype matrix multiplication app, REST execution time is about 200% better than SOAP execution approach. In case of energy consumption REST execution is about 250% better than SOAP execution approach.

Key Words: Execution Time, Battery Consumption, Soap Webservices, Rest Webservices

1. Introduction

Over the years, businesses interacted using ad hoc methods to get advantages of the basic network infrastructure [1]. however, Web services emerging application to application systematic approach for interaction which basis on existing protocols and XML standards. Web services can be defined as to expose the information system functionalities and make them available to public through Web technologies [2]. The term Web services in typical meaning is to serve something in Web and it defines the way of interacting the Web applications to each other using the WSDL (Web Services Description Language), UDDI (Universal Description Discovery and Integration), XML (Extensible Markup Language) and SOAP (Simple Object Access Protocol) which are open standards in a network protocol backbone. XML is used to distribute the computation over network by tagging the data, SOAP is a carrying protocol used to carry the data. WSDL describes the services available such as location of the services and methods of the services. UDDI is used for the discovery of suitable service of a particular task.

Web Services can also be defined as, an application accessible to other application over the Web. It is initially designed for application to application interaction regardless of the operating environment and developed where everyone agrees on some set of rules by which the interaction of Web Services takes place. As mobile devices by design carries limitations such as limited battery timing, slow processing, limited bandwidth [3, 4]. The similar concept can be implemented in mobile devices for computational offloading in order to address these limitations.

Amongst many other approaches of computational offloading one such approach is to utilize the idea of Web Services where the computational load is to delegate to remote servers for execution. Instead to have the clone of a mobile device or to delegate the whole application based on Virtual Machine Migrations, the mobile applications should be served as Web Services and the devices can directly approach to access those Web services without any management and overhead hurdles [5].

Each mobile application considered to be logically divided into two parts, one local part to reside in mobile device such as UI and the other should reside at remote server as a Web Service such as decoding or complex computational part. The existing mobile Web services are designed simply to serve the consumer on their mobile phones [6]. The intension of this work is to provide a self-configuring and self-provisioning service to address the complexity of traditional computational offloading techniques which are resources intensive and energy draining [7]. For the first time, we going to simulate and test the concept of Web services as a computational offloading solution for mobile computing in our lab environment. One closely related work done by [8], where they focused on to choose a best implantation structure for Web services amongst SOAP and REST in terms of services utilization. However, we differ here as our focus is to minimize the communication overhead by testing the developed prototype in our lab where we installed a middle layer next to remote server. It is to reduce the transmission distance to a single hop in order to address the limitations of mobile devices such as execution time delay and energy consumption through computational offloading.

To implement the concept of offloading through Web services, two types of load carrier protocols existed [9]. One SOAP (Simple Object Access Protocol), which is a standard object-oriented technology relying on the existing protocols such as SMTP and HTTP for messages negotiations and delegations. It uses XML format for the transmission of data through Web [10]. On the other hand, REST (Representational State Transfer Protocol) used for building up RESTful Web services. REST is categorized a resources oriented technology and developed by Roy Fielding [11] and it is an architectural style of World Wide Web which communicates over HTTP protocols. Uniform Interfaces which operates on URL resources, provides basis to RESTful Web. Both REST and SOAP are used to implement Web Services, however, each one has its own discrete features. We claim, that RESTful Web services carries a lightweight nature which are best fit for mobile computing and to address the limitations better than SOAP Web services.

Rest of the paper structured as, Section II provides the architectural comparison of SOAP and REST. Section III presenting the implementation methods of both the Services. Section IV discusses the results. Section V gives the future work and concludes the paper.

2. Architecture of Soap and Rest

A general overview of a SOA (Service Oriented Architecture) common both for SOAP and REST given in Fig. 1. WSDL a language used to create services description for a Service provider in the Service Directory. Services Directory which is used by the services providers to publish the services they offer. SOAP / REST is used as a protocol / architectural style by both the Services Providers and Consumers then to talk to the Services Directory. Next, based on the description given in Service Directory an XML message (XML a standard language used where both software can communicate in the same language to each other) is formed and then the Services Consumer requests to Service provider for permitting those particular Services. Both, REST and SOAP carrying their explicit pros and cons. For secure services utilization SOAP is considered the perfect solution, while for scalable and lightweight nature REST is the right choice [12].

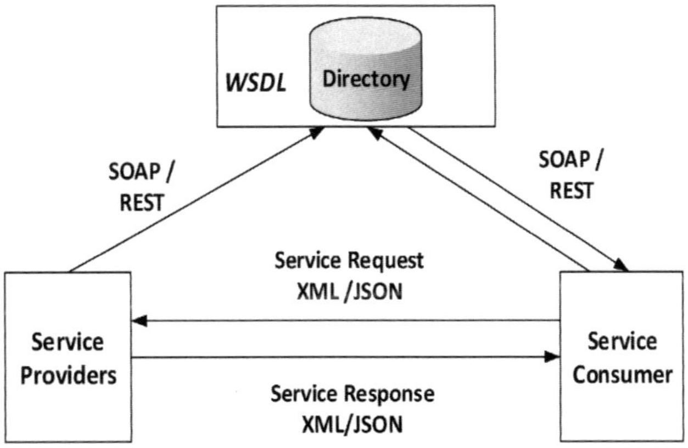

Figure 1: Service Oriented Architecture

Table 1. Presents a glance of comparison of both the REST and SOAP. Based on SOAP and REST, the web services having their own distinct features that turn the web service to a more or less suitable for a certain type of application and environment.

RESTful Web Services using a Unique URI (Uniform Resource Identifier) to describe the Resource. Using HTTP four basic operations such as GET, PUT, POST and DELETE are performed, while SOAP using a SOAP envelop which carriers the details of operations such as, starting of message, end of message, and the actual message in the form of XML.

A sample of SOAP message's envelope is:

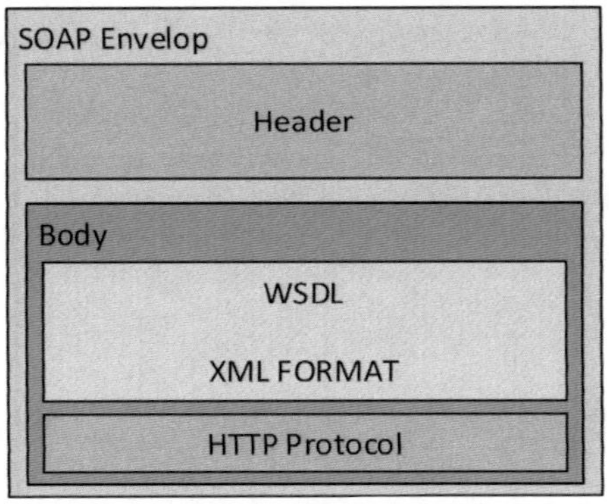

Figure 2: SOAP Envelope

Sample SOAP Message

```
<?xml version="1.0?>
<S:Envelope xmlns:S="http://
schemas.xmlsoap.org/soap/envelope/"
xmlns:SOAP-ENV="http://
schemas.xmlsoap.org/soap/envelope/">
.........
Messge information goes here
.........
</S:Envelope>
```

Figure 3: Sample SOAP Message

Every SOAP Envelope is the mandatory part of a SOAP message. It is the first root element which decides what should be sent to whom and what should be the rout of transmission. Each SOAP envelope having one head and one body element as shown in Fig. 2. The body of each SOAP envelope extricates between a SOAP request and SOAP Response. A typical SOAP request and SOAP response of a SOAP message shown in Fig 3. The message body contains the Name-Space of the Service and then two arguments arg0 and arg1 to the Server in a SOAP Request. Both the arguments contain 2, 2 which sent to the Service where both the arguments are taking as an input for a matrix multiplication service which generates two random matrices of size 2x2.

SOAP Request

```
<?xml version="1.0" encoding="UTF-
8"?><S:Envelope xmlns:S="http://
schemas.xmlsoap.org/soap/envelope/"
xmlns:SOAP-ENV="http://schemas.xmlsoap.org/
soap/envelope/">
    <SOAP-ENV:Header/>
    <S:Body>
        <ns2:MatrixMulti xmlns:ns2="http://
javabrains.ali.org/">
            <arg0>2</arg0>
            <arg1>2</arg1>

        </ns2:MatrixMulti>
    </S:Body>
</S:Envelope>
```

Figure 4: SOAP Request

SOAP Response

```
<?xml version="1.0" encoding="UTF-
8"?><S:Envelope xmlns:S="http://
schemas.xmlsoap.org/soap/envelope/"
xmlns:SOAP-ENV="http://
schemas.xmlsoap.org/soap/envelope/">
<SOAP-ENV:Header/>
<S:Body>
<ns2:MatrixMultiResponse
xmlns:ns2="http://javabrains.ali.org/">
<return>
<result>
<item>1625.0</item>
<item>2200.0</item>
</result>
<result>
<item>1232.0</item>
<item>2241.0</item>
</result>
</return>
</ns2:MatrixMultiResponse>
</S:Body>
</S:Envelope>
```

Figure 5: SOAP Response

REST	**SOAP**
Representational State Transfer	Simple Object Access Protocol
Suitable for Point to Point Communications	Designed for Distributed Computing
Minimize middle-wares, only HTTP needed	Needed middleware /tooling support
URL typical carrying the operation to be performed	Content of messages decides the operations
Security Issues	Secure
Not details description of WSDL services	Well-defined services structure
No Payload constraints	Payload need to comply with the SOAP policy
Built-in Error Handling	No Error Handling
Relying on HTTP transport model	SMTP and HTTP both are reliable layers for transmitting SOAP
Simple and Light	Complex and Heavy
An Architectural Style	A Protocol
SOAP can't use REST	REST can use SOAP as it is an architectural style and can use any protocol like SOAP, HTTP
Services Interface needed for exposing the business logic	URI needed to expose business logic
JAX-WS is the API in JAVA used	JAX-RS is Java API
Needed high bandwidth and resources	Needed less bandwidth and resources
Own Security defines	Inherit from underlying layer
Permit XML format only	Support different formats such as HTML, JSON, XML and Plain-text

TABLE 1. A COMPARISON OF SOAP AND REST

Both the randomly generated matrices are multiplied in the service and the resulted matrix is returned in the response as shown in Fig. 4. The XML tags <ns2:MatrixMulti.... </ns2:MatrixMulti> shows the SOAP request in Fig. 3 while <ns2:MatrixMultiResponse... </ns2:MatrixMultiResponse> shows the SOAP response in Fig 4. The resulted matrix should be of the same dimension 2x2. The returned Items are 1625, 2200 in the first row and 1232, 2241 are in the second row of the resulted matrix.

3. Implementation

The argument we test here that, offloading in mobile cloud using SOAP web services is heavyweight and resources intensive. SOAP needed more Turnaround Time (TT) and mobile battery consumption than REST. For that reason, to empirically test the intensiveness of SOAP against REST we planned to conduct the experiments in real lab environment.

In addition, a Server-Socket connection used to implement mobile Web serveries. There are five building blocks in Mobile Web Services: *Web Service-Servlet:* is used to deploy new services to mobile device and also invokes the demanded service. *HTTP Listener:* It has to listen to incoming requests coming through Server-Socket and accept incoming requests from client. *Request Handler:* It is responsible for processing the requests. The way this block work is different in both kind of frameworks. For instance, In SOAP-base framework the request handler unpacks the HTTP POST requests and extract the inside SOAP envelope then it dispatches the same envelope to message parser.

Conversely, request handler for RESTful Web Services extract directly the HTTP request and send it to message parser. Parser Module: it gets the required information of a Web Service such as name, URL and parameters of a Service. The extracted information is then send to Service-Servlet. This is performed different in two frameworks. In SOAP-based Web services, the parser de- serializes the object and by using KSOAP2 and kXML2 (Open source APIs for SOAP parsing) maps the data types to Java Objects. In RESTful Web Services, we used to create our own manual string manipulator which do Parsing. Response Composer: It interpret the results and then send it back to client.

Experiment Setup:

The experimental emulation environment planned. The setup composed of remote surrogate machine running Microsoft Windows 10 Professional 32-bit operating system with Intel® core(TM) i7-3770 @ 3.40GHz speed and 4.0 GB RAM capacity and a Mobile client device Samsung Galaxy A5 runs Android 5.0.2(Lollipop) OS with Quad Core 1.2 GHz Cotex-A53 Processor, 16GB memory 2GB RAM and 2300mAh battery. A D-Link wireless Wi-Fi modem / Access Point with a physical layer data rates 54 Mbps used to connect the remote server machine and mobile devices. Mobile device accesses the wireless network via Wi-Fi wireless network connection of radio type 802.11g, with the available physical layer data rates of 54Mbp.

Two components of the proposed framework developed. One for Client side running at mobile device while the other runs at surrogate servers. The Server-side application (Web service) deployed by configuring a Glassfish server at surrogate using Web Services Application tools in Eclipse. Also, Android Developers (Java based Android Software development tool kit) deployed for the development of client-side application in Intellij IDE 15 Community environment. (Android Debug Bridge) ADB plugin used to develop the prototype android application.

Two porotype applications named SOAP-app and REST-app developed which required to get two numeric integer values as arguments, pass the arguments to Web Server. Web Server randomly generates two matrices of the same input size and multiply both. The resulted matrix gets back to client device. The execution time and battery consumption tested for both. A stop watches Time Left used to read TT while Power Tutor to read the energy consumption.

4. Result and Discussion

As the matrices multiplication is a computational intensive task, a porotype developed to multiply two randomly generated matrices ranged between 300*300-400*400. Each computational intensity varies from the previous one with an addition of 10. The total 10 different computational intensities ran. Though, different parameters such as, fluctuating bandwidth and loss of network signals, remote server load and so on, may affect the result of executing a task at remote servers, therefore each intensity iterated for 5 times in order to eliminate ambiguity in the observations. For further validation mean value of the collected 5 sample values calculated. The execution time while executed the prototype through SOAP web service is shown in Figure 6.

The execution of lowest intensity 300*300 executed in approx. 30 seconds while the highest intensity 400*400 executed in 140 seconds. Similarly, the execution of all the intensities through REST web services is as shown in Figure 7, the lowest intensity 300*300 executed in approx. 20 seconds while the highest intensity in 38 seconds.

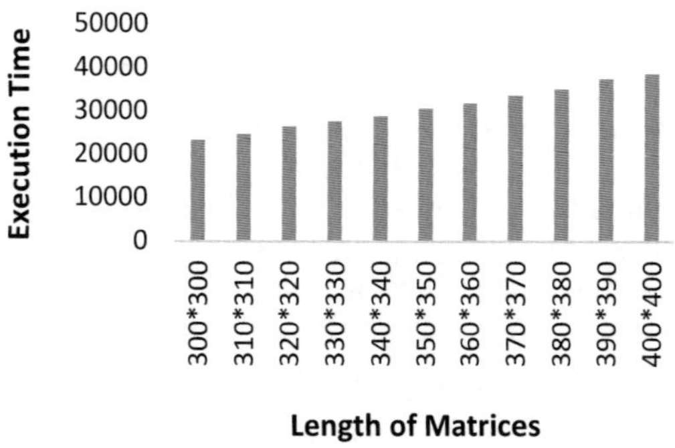

Figure 6: Execution Time of SOAP Web services

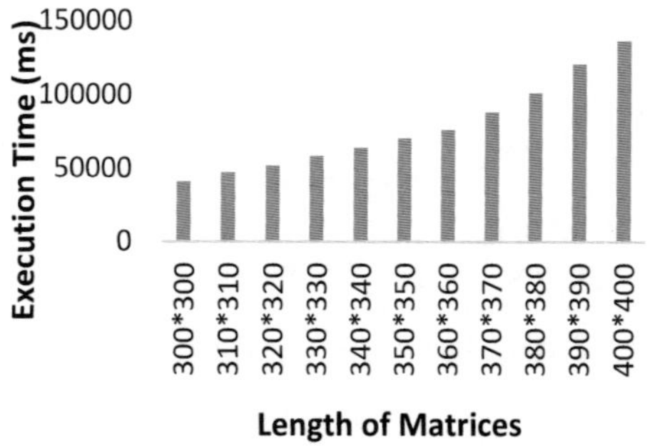

Figure 7: Execution Time of REST Web services

The comparison of both the services as shown in Figure 8 illustrates that for the same data to transmit and execute with the same lab test using same computing devices needed more than double execution time to execute by SOAP web services as compare to REST web services.

The execution time of SOAP based Web services is highest than the RESTful Web services and it is due the heavyweight process of un-wrapping the SOAP envelope for each message request and then the SOAP object de-serializes and mapping the data types of actual messages (XML based) into the Java objects. Conversely, the execution time of RESTful Web services is almost half of the execution time of SOAP web services and the reason is, the parsing of heavyweight process eliminated by a lightweight parsing procedure and to extract the hidden messages. As, REST no need to use additional libraries for parsing SOAP response and utilizes less resources compare to SOAP. Also, SOAP needed to have a contract between remote system and client device in order to make the communication secure and effective it need to define schemes and implement policies to establish a secure and smooth flow of data.

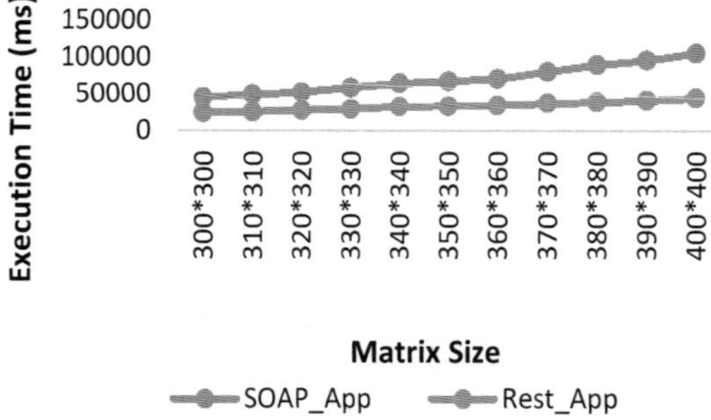

Figure 8: Comparison of ET between SOAP and REST Web services

Similarly, the energy consumption cost of the same input arguments of executing through SOAP web services is shown in Figure 9. The energy consumption cost in Joule (J) of the lowest computational intensity 300*300 is approx. 30J while the same for the highest intensity 400*400 is 60J.

On the other hand, the execution of computational intensities through REST web services gives the results as shown in Figure 10. The energy consumption cost of the lowest intensity 300*300 is approx. 10J while the same of highest intensity 400*400 is approx. 20J. The comparison of both the services as given in Figure 11 shows that executing the same intensity by the both the services, the SOAP web services hit the battery more compare to the REST services.

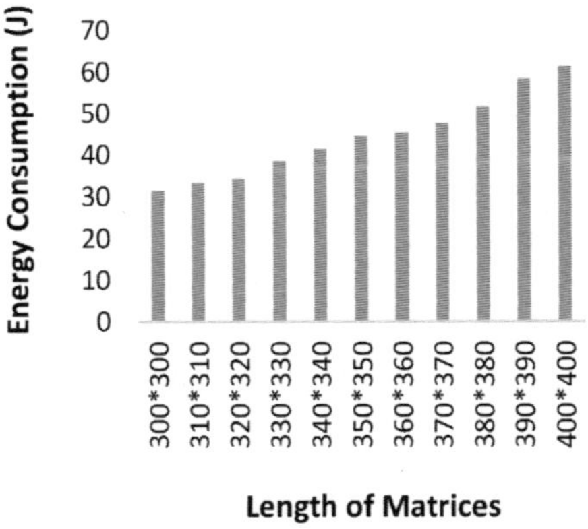

Figure 9: Energy Consumption Cost of SOAP Web Services

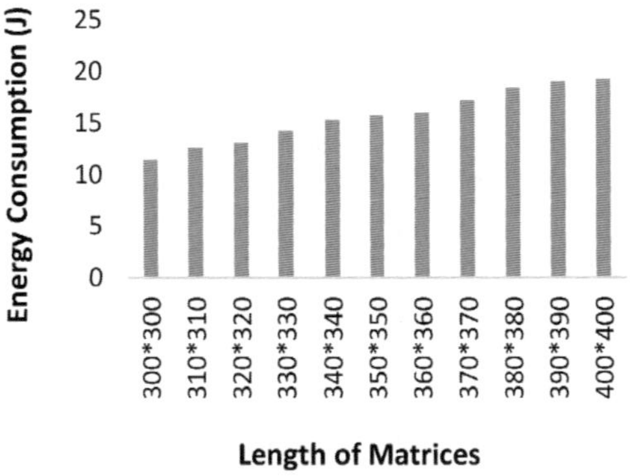

Figure 10: Energy Consumption Cost of REST Web Services

By executing the same intensity in order to identify the energy intensive service in two, it is clear from Figure 11 that SOAP Web Services drain power of mobile device more than double compare to REST services. The energy consumption of lowest intensity 300*300 of REST Web services starting from approx. 10J and reaches up to 18J, while the same of SOAP Web Services starting from 30J of the lowest computational intensity and reaches up to 60J for the highest one.

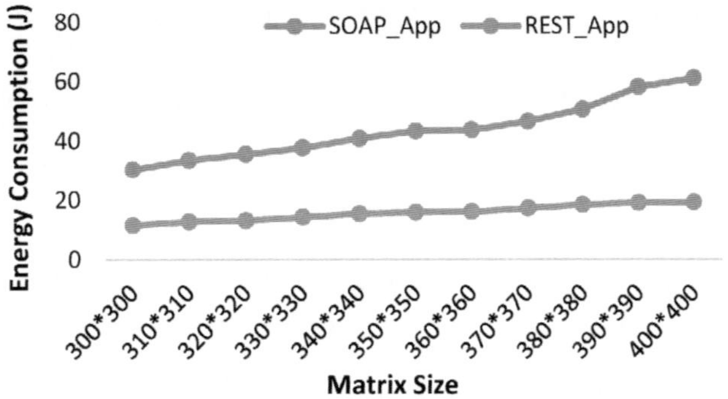

Figure 11: Comparison of EC between SOAP and REST Web Service

5. Conclusion and Future Work

Web utilities, where application to application interactions takes place are used to let electronic devices communicate with each other and initially Web Services designed to support interoperability of machine-to-machine interactions using World Wide Web. Combining Web services with mobile computing drives towards an approach where the limitations of mobile devices can be addresses. Web services basis on two types of technologies, SOAP implemented and REST implemented. Both, SOAP and REST carries their unique features which makes them applicable for different scenarios. In case of mobile computing, the intension of this work is to use Web services and delegate the computational intensive tasks to remote servers utilizing both type of Web Services, in order to identify that which service is more lightweight and suitable for mobile computing, as a result to minimize resources consumption at mobile device.

To test the effective approach amongst SOAP and RESTful web services, we conducted the experiments in real scenario. Prototype applications SOAP-App and REST-App developed to empirically evaluate the intensity of each Web service in terms of execution time and energy consumption with mobile computing. The prototype applications consist of two components, one component run local whereas the next component run at remote server. The local component consist of UI which get input from users and sends to remote part of the app to execute. The experimental data collected in REST and SOAP both the approaches separately with 10 different computational intensities (300*300-400*400). For validation purposes, each intensity in both the scenarios iterated 5 times and a mean value calculated of execution time and energy consumption. The results show that due to heavyweight parsing procedure of SOAP Web services REST Web services are more effective to use in mobile computing. The execution time recorded in our experiments of REST Web services is nearly 200% better than SOAP while the energy consumption cost of executing same application of both scenarios, REST Web services 250% effective than SOAP Web services.

It is intended to explore more feature of both the Web services and consider maximum parameter which can affect mobile computing with Web services will be publish in the next extended article with complete details. Further, the experiments conducted with one single model of mobile device Samsung Galaxy A5, which will be extended to any different model of mobile device for expanding the scope of the research.

ACKNOWLEDGMENT

This work is fully funded by University Malaysia Pahang under Short Term Research Grant, ref: RDU 140391. I acknowledge the support of the Faculty of Computer Systems & Software Engineering, University Malaysia Pahang by providing a suitable environment for this research.

REFERENCES

[1] F. Curbera, M. Duftler, R. Khalaf, W. Nagy, N. Mukhi, and S. Weerawarana, "Unraveling the Web services web: an introduction to SOAP, WSDL, and UDDI," *IEEE Internet computing,* vol. 6, pp. 86-93, 2002.

[2] G. Alonso, F. Casati, H. Kuno, and V. Machiraju, "Web services," in *Web Services*, ed: Springer, 2004, pp. 123-149.

[3] M. Ali, J. M. Zain, M. F. Zolkipli, and G. Badshah, "Mobile cloud computing & mobile battery augmentation techniques: A survey," in *Research and Development (SCOReD), 2014 IEEE Student Conference on*, 2014, pp. 1-6.

[4] M. Ali, J. M. Zain, M. F. Zolkipli, and G. Badshah, "Mobile cloud computing & mobile's battery efficiency approaches: A Review," *Journal of Theoretical and Applied Information Technology,* vol. 79, p. 153, 2015.

[5] S. Anwar, Z. Inayat, M. F. Zolkipli, J. M. Zain, A. Gani, N. B. Anuar, *et al.*, "Cross-VM Cache-based Side Channel Attacks and Proposed Prevention Mechanisms: A survey," *Journal of Network and Computer Applications,* vol. 93, pp. 259-279, 1/09/2017 2017.

[6] P. Farley and M. Capp, "Mobile web services," *BT Technology Journal,* vol. 23, pp. 202-213, 2005.

[7] M. Shiraz, M. Sookhak, A. Gani, and S. A. A. Shah, "A study on the critical analysis of computational offloading frameworks for mobile cloud computing," *Journal of Network and Computer Applications,* vol. 47, pp. 47-60, 2015.

[8] F. AlShahwan and K. Moessner, "Providing soap web services and restful web services from mobile hosts," in *Internet and Web Applications and Services (ICIW), 2010 Fifth International Conference on*, 2010, pp. 174-179.

IOP Conf. Series: Journal of Physics: Conf. Series **1018** (2018) 012005 doi:10.1088/1742-6596/1018/1/012005

[9] M. Lanthaler and C. Gütl, "Towards a RESTful service ecosystem," in *Digital Ecosystems and Technologies (DEST), 2010 4th IEEE International Conference on*, 2010, pp. 209-214.
[10] K. Devaram and D. Andresen, "SOAP optimization via parameterized client-side caching," in *Proceedings of the IASTED International Conference on Parallel and Distributed Computing and Systems (PDCS 2003)*, 2003, pp. 785-790.
[11] R. T. Fielding, "Architectural styles and the design of network-based software architectures," University of California, Irvine, 2000.
[12] C. A. Binildas, M. Barai, and V. Caselli, "Service Oriented Architecture with Java," *Using SOA and web services to build powerful Java applications. Birmingham,* 2008.

IOP Conf. Series: Journal of Physics: Conf. Series **1018** (2018) 012006 doi:10.1088/1742-6596/1018/1/012006

A Review Study on Cloud Computing Issues

Qusay Kanaan Kadhim [1, 3], Robiah Yusof [1], Hamid Sadeq Mahdi[3], Sayed Samer Ali Al-shami [2], Siti Rahayu Selamat[1]

[1]Faculty of Information and Communication Technology,
[2] Institute of Technology Management and Entrepreneurship
[1, 2] Universiti Teknikal Malaysia Melaka (UTeM)
Hang Tuah Jaya, 76100 Durian Tunggal, Melaka, Malaysia.
[3] Department of Computer Science, Al Yarmouk University College, Iraq.
[3] University of Diyala / Basic Education College / Computer Science Department, Iraq.
corresponding author's Email: qusaykanaan@utem.edu.my

Abstract. Cloud computing is the most promising current implementation of utility computing in the business world, because it provides some key features over classic utility computing, such as elasticity to allow clients dynamically scale-up and scale-down the resources in execution time. Nevertheless, cloud computing is still in its premature stage and experiences lack of standardization. The security issues are the main challenges to cloud computing adoption. Thus, critical industries such as government organizations (ministries) are reluctant to trust cloud computing due to the fear of losing their sensitive data, as it resides on the cloud with no knowledge of data location and lack of transparency of Cloud Service Providers (CSPs) mechanisms used to secure their data and applications which have created a barrier against adopting this agile computing paradigm. This study aims to review and classify the issues that surround the implementation of cloud computing which a hot area that needs to be addressed by future research.

Keywords: Cloud, Computing, Security, Issues

1. Introduction

Cloud computing becomes a promising networking for infrastructure pattern which can deploy large-scale application in a cost-effective method. It is defined as "applications delivered as services over the Internet and the hardware and systems software in the data centers that provide those services [1]. Recently, cloud computing has been widely adopted by the industry and organizations due to its usability and simple cloud of services oriented models. The number of cloud users can access the cloud of services to keep on the increasing daily and safe systems in the cloud computing environments. Cloud computing technology plays an important role in academic and industry organizations. The business processes are composed and implemented in the distributed loosely coupled environments and the composite of services which includes more services and thus the cloud of services will be connected by varies patterns and approaches. Cloud computing is providing organizations to use shared data storage and cloud resources. It is better than to develop with the own platforms. Further, cloud computing provides companies to have a data flexible, secure system, and cost-effective cloud infrastructure [2]. Additionally, the cloud computing can provide on demand dynamically scalable virtualized cloud resources via the web of internet. indeed, the cloud computing has not only changed the way of providing cloud services but influenced the way of application development, which helps companies to save IT resources during the lifecycle and shorten application development time [3].

Despite, the advantages of the cloud computing, it still surrounded by several issues that are associated with security management [4] includes lack of trust in data security and privacy by users, organizational inertia, loss of governance, and uncertain provider's compliance [5]. The security issue became extra complex under the cloud model as new scopes have arrived into the problem scope associated to the model data security [6], users' privacy [7], network security, platform and infrastructure issues. Recently, studies from various disciplines emphasized to the importance of cloud computing security management in all areas of application to mitigate those issues. The new version of cloud security management consists of the processes and methods that are useful to reduce cloud security issues. It also includes characteristics like on-demand service provision, virtualization and virtual data centers, and high flexibility access to data on cloud storage and release of service provision like storage, network, cloud applications, servers and its cloud services. Indeed, they present the conceptual model for cloud security that involved components such as data privacy, legally and standard, policy, compliance and regulatory issues of government organizations. Due to the fact that there are many types of security issues, this study reviewed many types of security issues. This study also classified the sub-issues of cloud computing under structured groups which helps future research to explore the related solution. To classify the security issues in the government, this study interviewed the managers of information technology department in 23 of Iraqi government departments and they classified the security issues under five main groups which are: - mobility and cloud government application security issues, cloud security services and application issues, cloud security data, cloud network security issues and issues for cloud security platform and infrastructure.

2. Literature Review

2.1. Cloud Computing Issues

There are many cloud security issues appear in different type of technologies which include networks, databases, operating systems, virtualization, resource scheduling, transaction management, concurrency control and memory management are used in cloud computing [8]. The cloud computing is designed as computing utility. The majority of individuals and enterprises use to migrate their work into the cloud, workloads which became more heterogeneous because the cloud computing resources are much more heterogeneous as cloud providers constantly scale or update the clusters with new generations of machines. For example, the government organizations run applications and data transfer in their own the private cloud and then transform it to the public cloud. Nevertheless, there are many security issues exist in the cloud computing technology that threaten the data credibility and confidently. This emphasis to the important to design a cloud computing security with relevant standards and policies to protect the users. Despite, there are major efforts made to design effective cloud computing, the cloud service security is still facing new business and management problems arising from virtualization, multi-tenancy in the cloud, and data encryption technology, trusted cloud and cloud data sensitive confidentiality issues [9]. The government and non-government organizations are struggling to identify the features of cloud security issues and then build and prepare plan that can help them to make appropriate decision toward a successfully adoption of cloud computing technology projects for organizations. There are several key cloud security challenges within the cloud computing environment as shown in Figure , which include mobility and application security issues [10], cloud security services and applications issues, cloud security data issues [11], cloud network security issues [12] and cloud security platform and infrastructure issues [13]

2.1.1. Mobility and Cloud Government Application security issues

Despite the growth usage of mobile computing, exploiting its full potential advantages is difficult due to its inherent problems such as secure resource, frequent disconnections, and mobility [14].Data in the cloud is typically in a shared environment alongside with data from other customers. Encryption is critical to protect data sensitive confidentiality and privacy of the data while in transiting and in cloud storage. Legally, third-party cloud service providers (CSPs) and their customer organizations are

IOP Conf. Series: Journal of Physics: Conf. Series **1018** (2018) 012006 doi:10.1088/1742-6596/1018/1/012006

distinct enterprises. It the CSP fails in its responsibilities, it could have legal liability implications for the CSP's customer organizations. In contrast, if a cloud customer organization fails in its responsibilities, it is less likely to be exposure to legal implications of the CSP [15]. There are some responsibilities for the organizations such as flexibility access issue of cloud providers, protect sensitive data in the cloud computing, understand legally and standards issues, software development life cycle management, portability and interoperability, and cloud platform reliability and latency. The policy is a foundational issue that is related to legal definition and organizational charter to facilitate and guide the establishment of the vision, missions, responsibilities, and authorities of major actors in cloud computing organization. Open Security Architecture (OSA) provides frameworks that are easily integrated into software applications for the security architecture community. Its patterns are based on schematics that show the data traffic flow control for secure cloud computing and particular implementation with policies implemented at each step for the cloud security issues. The cloud mobile applications can connect and request services hosted on a remote cloud computing by interfaces [18]. However, mobile Web services need to consider additional constraints other than standard Web services: frequent loss of connectivity, low computational resources, and low bandwidth [37]. In this section, the mobility and cloud government application security issues are discussed.

Table 1 Issues for Mobility and Cloud Government Application Security Based on Studies

Mobility and Cloud Government Application security issues	Related works
Lack of Standards, Legally, and Policy	[16]–[19].
Loss of Security Governance	[20]–[22].
Malicious Insider Threats in the Cloud Computing	[23], [24][25]
Cloud Computing Regulatory Requirements and Cloud Compliance Challenges	[22], [26]–[29].
	[30][31]
Cloud Computing Portability and Interoperability	
Biometric Security System for Cloud Computing Environment	[32][33][34][35]

2.1.2. Cloud Security Services and Application Issues

Service and application level issues relate to the security factors concerned with performance measurements of the cloud computing system, and the quality of service and cloud service level agreement [41]. For example, in what ways do mobile cloud computing systems ensure data of availability; what are the fault- tolerance (FT) mechanisms employed to ensure smooth execution and uninterrupted service [42]. Therefore, cloud of services able to extend dynamically to meet user on demand and requirements. Additionally, capabilities of the cloud can be rapidly and elastically increased to meet immediate demand and scaled down to release unused resources. However, monitoring should be done by the cloud service provider frequently to ensure that the cloud is secure and not infected by the malicious code that have been uploaded to the cloud by the hackers or attackers with the purpose of stealing sensitive data or even damaging the information of users. Further, Table 0, illustrates the issues for cloud security services and application based on studies bellow.

Table 2. Issues for Cloud Security Services and Application Based on Studies

Cloud Security Services and Application Issues	Related works
Cloud Service Level Agreement (SLAs) and Quality of	[36]–[45].
Service	[22], [43], [46]–[54].
Trusted for Cloud Services	[2], [16], [24], [43], [48], [52], [54].
Access Control in Cloud Computing Environment	[23], [25], [27], [28], [55].
Security of Cloud Interfaces and API	[38], [48], [58]–[62].
Availability of Cloud Data	

2.1.3. Cloud Security Data

Generally, the data in cloud computing belongs to different owners in the cloud computing resources which must be trusted. Therefore, unauthorized users should be forbidden from that data or information [52], [54], [59]. In trusted and cloud data sensitive confidentiality refers to original data that must keep in a password protected by data management systems with security guard services in the cloud computing environments. All cloud data are entered, stored and backed-up in a password-protected by the management of data in the cloud computing [63]. Besides, the cloud data storage is a model of data storage in which the integrity data is stored in logical pools. It allows cloud users to store their data in a remote server to get rid of expensive local storage and managing brand cost and then flexibility access data of interest anytime and anywhere [64].

Table 3: Issues for Cloud Security Data Based on Studies

Cloud Security Data Security	**Related works**
Cloud Data Privacy Security	[16], [17], [22], [25], [27], [31], [58], [65].
Data Protection in Cloud Computing Environments	[2], [54], [55], [47].
Cloud Data Confidentiality Issues	[48], [49], [52], [54].
Cloud Data Limitations and Segregation	[31], [62], [60]–[61].
Cloud Data Integrity	[16], [24], [27], [52], [54], [59].
Cloud Data Eavesdropping Attack and Leakage	[23], [33], [39], [55], [57].

2.1.4. Cloud Network Security Issues

Cloud network security is one of the network security issues. Cloud Computing permits ever-present, convenient, on-demand network access to a shared pool of configurable networks that can be quickly provisioned and free with negligible management effort or service provider communication [67]. Due to the fact that the Cloud Computing embodies a comparatively new computing model, there is an important deal of vagueness about how security at network can be attained and how applications security is progressed to Cloud Computing [62]. The cloud network issues are the higher response time of nodes while performing data communication through co-operative caching [8].

Table 4: Issues for Cloud Network Security Based on Studies

Cloud Network Security issues	**Related works**
Detection and Recovery	[2], [15], [31], [52], [55], [59].
Flow Control for Secure Cloud Computing	[2], [27], [43], [58].
Cloud Account or Cloud Service Hijacking	[19], [40], [63], [64] .
Cloud Network Traffic Analysis and control	[21], [25], [45], [65].
Bandwidth Cost in the Cloud	[26], [50], [51], [58]].
Distributed Denial of Service (DDoS) Attacks for the Cloud	[2], [15], [47], [66].

2.1.5. Cloud Security Platform and Infrastructure Issues

Security of the cloud infrastructure relies on trusted cloud and cryptography. In addition, no standard service contract exists that covers the ranges of cloud services available and the needs of different organizations. Beside that the cloud computing technology allow to design and implemented a real time alert system on top of the cloud infrastructure [44]. However, cloud computing offers an elastic infrastructure that Cloud Management Agents can use to obtain streaming resources that match the demand. Also, multi cloud providers support different platforms and offer constantly changing packages of capabilities. Further, infrastructure security is the basis of cloud computing security, mainly in the cloud for the upper layer of security services to provide security, infrastructure security by hardware and software security, can be a variety of intrusion defense, redundant backup of data, intrusion detection and prevention in network security. Cloud infrastructure security issues the risk associated end-user's concern and is also the focus this work's research direction. Therefore, the centralized security solution for insecurity cloud is proposed and the scenarios of this system and methods after that constructed.

Table 5: Issues for Cloud Security Platform and Infrastructure Based on Studies

Cloud Security Platform and Infrastructure Issues	**References**
Cloud Platform Reliability and Latency	[31], [65], [71]–[68].
The multi-tenancy in the Cloud Scalability and Capability in the Cloud	[15], [24], [27], .

3. Cloud Security Issue Factors

Many cloud security issues can obtain from different technologies that including networks, databases, operating systems, virtualization, resource scheduling, transaction management, concurrency control and memory management are used in cloud computing [8]. Additionally, cloud computing is designed as computing as a utility. Customers rent computing resources in the cloud to complete their work. Then to ensure the quality of service (QoS) requirements defined by customers and guarantee the resource utilization in the cloud datacenters, effective resource management systems should be considered. Therefore, more individuals and enterprises migrating their work into the cloud, workloads in the cloud become more heterogeneous. Meanwhile, cloud computing resources are much more heterogeneous as cloud providers constantly scale or update the clusters with new generations of machines [69]. However, in reality government organizations run applications and data transfer in their own the private cloud and then transmute it to the public cloud. While there are many security issues exist in the cloud computing technology , cloud security should design relevant standards and policies as soon as possible [43].

Cloud computing is a new emerging technology, which every organization these days wants to adapt for its business for more profitability, interoperability, capability, and scalability. This network communication defined cloud computing, highlighted all the cloud service models likes public, private, hybrid and community cloud computing. In addition to information security risks under traditional IT architecture, cloud service security is still facing new business and management risk arising from virtualization, multi-tenancy in the cloud, and data encryption technology, trusted cloud and cloud data sensitive confidentiality issues [9]. However, cloud computing is changing the current IT delivery model for services. Benefits for business and IT include reduced costs, scalability, flexibility access to data on cloud storage, capacity utilization, higher efficiencies, performance, and mobility [40]. The enterprise will be able to identify the features of cloud security issues and then build and prepare plan that can help them to make appropriate decision toward a successfully adoption of cloud computing technology projects for organizations [60]. Generally, cloud computing's issues can bring negative effects on any companies or organizations, therefore an effective risk management is needed to balance the operational and financial cost as well as proactive actions to secure data, network, platform information systems and technologies [70]. According to [53], understanding the true potential of mobile cloud computing and identifying issues with mobile cloud security, privacy, feasibility and accessibility remain a major concern for both the customers and the enterprises [57]. Therefore, the presence of cloud security issues and challenges have critical influence on the success of cloud computing systems. Thus, it is critical to identify and classify control various issues during cloud computing environments by using controlling mitigation techniques those security issues, the success rate of cloud computing systems could be increased. Cloud security, cloud data privacy, feasibility and accessibility remain a major concern for both the users and the enterprises [56]. There are several key cloud security challenges within the cloud environment such as [28]: Key stores that must be protected in data storage, detection and recovery and in backup. Improper key storage may lead to encryption data. Flexibility accesses to key data storage have to be limited to the authorized personnel who require the individual keys. These keys ought to be under policies governing them [45].

4. Conclusion

Cloud computing is a new emerging technology, which every organization these days adapt it to facilitate the flexibility of their businesses in terms data storage, exchange, transform which enable them to upgrade their profitability, interoperability, capability, and scalability. This network communication defined cloud computing, highlighted all the cloud service models likes' public, private, hybrid and community cloud computing. The cloud computing is changing the current IT delivery model for services. Benefits for business and IT include reduced costs, scalability, and flexibility access to data on cloud storage, capacity utilization, higher efficiencies, performance, and mobility. Despite, the advantages of the cloud computing, it still surrounded by several issues that are associated with security management, includes lack of trust in data security and privacy by users, organizational inertia, loss of governance, and uncertain provider's compliance. The security issue became extra complex under the cloud model as new scopes have arrived into the problem scope associated to the model data security, users' privacy network security, and platform and infrastructure issues. This study was designed to highlight the cloud computing security issues. The finding of this study emphasises that there are five main issues associated with cloud computing implementation which are Mobility and Cloud Government Application security issues, Cloud Security Services and Application, Cloud Security data, cloud network security issues and cloud security platform and infrastructure issues. These issues form an open room for future research to fill up security issues gap through providing either technical approach or empirical model to mitigate these issues.

Acknowledgements

The authors would like to thank INSFORNET Research Group of Universiti Teknikal Malaysia Melaka for supporting this research

References

[1] M. Armbrust, A. Fox, R. Griffith, A. Joseph, R. Katz, A. Konwinski, G. Lee, D. Patterson, A. Rabkin, and I. Stoica, "A View of Cloud Computing," *Commun. ACM*, vol. 53, no. 4, pp. 50–58, 2010.

[2] K. Jakimoski, "Security Techniques for Protecting Data in Cloud Computing," *Int. J. Grid Distrib. Comput.*, vol. 9, no. 1, pp. 49–56, 2016.

[3] Z. Hong-lie, L. Xin, L. I. U. Yan-ju, and L. Cheng, "Research on Cloud Resource Section Method for the Multi-layer Ontology," *Int. J. Grid Distrib. Comput.*, vol. 9, no. 1, pp. 193–200, 2016.

[4] A. N. Khan, M. L. Mat Kiah, S. U. Khan, and S. A. Madani, "Towards secure mobile cloud computing: A survey," *Futur. Gener. Comput. Syst.*, vol. 29, no. 5, pp. 1278–1299, 2013.

[5] A. M.-H. Kuo, "Opportunities and Challenges of Cloud Computing to Improve Health Care Services," *J. Med. Internet Res.*, vol. 13, no. 3, p. e67, 2011.

[6] L. M. Kaufman, "Data security in the world of cloud computing," *IEEE Secur. Priv.*, vol. 7, no. 4, pp. 61–64, 2009.

[7] M. B. Mollah, M. A. K. Azad, and A. Vasilakos, "Security and privacy challenges in mobile cloud computing: Survey and way ahead," *Journal of Network and Computer Applications*, vol. 84. pp. 38–54, 2017.

[8] D. Sarddar, P. Sen, and M. K. Sanyal, "Central Controller Framework for Mobile Cloud Computing," *Int. J. Grid Distrib. Comput.*, vol. 9, no. 4, pp. 233–240, 2016.

[9] Z. Gao, Y. Li, H. Tang, and Z. Zhu, "Management Process Based Cloud Service," in *International Conference on Cyberspace Technology (CCT 2013)*, 2013, pp. 278–281.

[10] A. Botta, W. de Donato, V. Persico, and A. Pescapé., "Integration of cloud computing and

internet of things: a survey.," *Futur. Gener. Comput. Syst.*, vol. 56, p. 684–700, 2016.

[11] D. A. B. Fernandes, L. F. B. Soares, J. V. Gomes, M. M. Freire, and P. R. M. Inácio, "Security issues in cloud environments: a survey," *Int. J. Inf. Secur.*, vol. 13, no. 2, pp. 113–170, 2014.

[12] V. Inukollu, S. Arsi, and S. Ravuri, "Security Issues Associated With Big Data in Cloud Computing," *Int. J. Netw. Secur. Its Appl.*, vol. 6, no. 3, pp. 45–56, 2014.

[13] H. Rasheed, "Data and infrastructure security auditing in cloud computing environments," *Int. J. Inf. Manage.*, vol. 34, no. 3, pp. 364–368, 2014.

[14] N. Fernando, S. W. Loke, and W. Rahayu, "Mobile cloud computing: A survey," *Futur. Gener. Comput. Syst.*, vol. 29, no. 1, pp. 84–106, 2013.

[15] C. LLP, W. Chan, E. Leung, and H. Pili, "Enterprise Risk Management for Cloud Computing," 2012.

[16] A. Tuli, N. Hasteer, M. Sharma, and A. Bansal, "Exploring Challenges in Mobile Cloud Computing: An Overview," *Confluence 2013: The Next Generation Information Technology Summit (4th International Conference)*. p. 6, 2013.

[17] NSTAC, "NSTAC Report to the President on Cloud Computing," 2012.

[18] F. Al-anzi, S. Yadav, and J. Soni, "Cloud Computing: Security Model Comprising Governance, Risk Management and Compliance," in *2014 International Conference on Data Mining and Intelligent Computing (ICDMIC),* 2014, pp. 1–6.

[19] R. Matt, "Cybersecurity and Cloud Computing in the Health Care and Energy Sectors: Perception and Reality of Risk Management," 2013.

[20] E. Takamura, C. Gomez-rosa, K. Mangum, and F. Wasiak, "MAVEN Information Security Governance , Risk Management , and Compliance (GRC): Lessons Learned," in *2014 IEEE Aerospace Conference*, 2014, pp. 1–12.

[21] J. Adjei, "Explaining The Role of Trust in Cloud Service Acquisition," in *2014 2nd IEEE International Conference on Mobile Cloud Computing, Services, and Engineering*, 2014, pp. 283–288.

[22] S.-T. Lai and F.-Y. Leu, "A Security Threats Measurement Model for Reducing Cloud Computing Security Risk," in *2015 9th International Conference on Innovative Mobile and Internet Services in Ubiquitous Computing*, 2015, pp. 414–419.

[23] P. Anand, J. Ryoo, H. Kim, and E. Kim, "Threat Assessment in the Cloud Environment – A Quantitative Approach for Security Pattern Selection," in *IMCOM '16*, 2016, p. 8.

[24] E. Cayirci, "Modeling and Simulation as A Cloud Service: A Survey," in *Proceedings of the 2013 Winter Simulation Conference*, 2013, pp. 389–400.

[25] A. Michalas, N. Paladi, and C. Gehrmann, "Security Aspects of e-Health Systems Migration to the Cloud," in *2014 IEEE 16th International Conference on e-Health Networking, Applications and Services (Healthcom) Security*, 2014, pp. 212–218.

[26] P. Hazarika, V. Baliga, and S. Tolety, "The Mobile-Cloud Computing (MCC) Roadblocks," in *2014 Eleventh International Conference on Wireless and Optical Communications Networks (WOCN)*, 2014, pp. 1–5.

[27] M. Bamiah, S. Brohi, and S. Chuprat, "Cloud Implementation Security Challenges," in *Proceedings of 20l21ntemational of Cloud Computing, Technologies, Applications & Management*, 2012, pp. 174–178.

[28] F. Al-Musawi, A. H. Al-Badi, and S. Ali, "A Road Map to Risk Management Framework for Successful Implementation of Cloud Computing in Oman," in *2015 International Conference on Intelligent Networking and Collaborative Systems*, 2015, pp. 417–422.

[29] J. K. Ganlea, K. Afriyie, and A. Y. Segbefia, "Microcredit: Empowerment and Disempowerment of Rural Women in Ghana," *World Dev.*, p. Pages 335–345, 2015.

[30] E. Aruna, A. Shri, and A. Lakkshmanan, "Security Concerns and Risk at Different Levels in Cloud Computing," in *2013 International Conference on Green Computing, Communication and Conservation of Energy (ICGCE)*, 2013, pp. 743–746.

[31] B. Shanthini and S. Swamynathan, "Genetic-based biometric security system for wireless sensor-based health care systems," in *Proceedings of the 2012 International Conference on Recent Advances in Computing and Software Systems, RACSS 2012*, 2012, pp. 180–184.

[32] G. Ahammed, R. Banu, and N. Fathima, "An Approach to Secure Communication in IoT (Internet of Things)," in *CONFERENCE ON INTERNET OF THINGS*, 2016, no. February, p. 315.

[33] C. Klein, "Cloudy Confidentiality : Clinical and Legal Implications of Cloud Computing in Health Care," *Anal. Comment.*, vol. 39, no. 4, pp. 571–578, 2011.

[34] A. Wadhawan and A. Bhatia, "Neural Network Based Intelligent Retrieval System for Verifying Dynamic Signatures," *Int. J. Adv. Sci. Technol.*, vol. 83, no. 2015, pp. 27–40, 2015.

[35] M. Alhomidi and M. Reed, "Security Risk Analysis as a Service," in *2013 8th International Conference for Internet Technology and Secured Transactions, ICITST 2013*, 2013, pp. 156–161.

[36] A. Khan, M. Fayaz, A. S. Shah, and F. Wahid, "Critical Analysis of Cloud Computing Software Development Process Models," *Int. J. Softw. Eng. Its Appl.*, vol. 10, no. 11, pp. 451–466, 2016.

[37] E. Arianyan, M. Ahmadi, and D. Maleki, "A Novel Taxonomy and Comparison Method for Ranking Cloud Computing Software Products," *Int. J. Grid Distrib. Comput.*, vol. 9, no. 3, pp. 173–190, 2016.

[38] M. Carroll, A. Merwe, and P. Kotzé, "Secure Cloud Computing: Benefits, Risks and Controls," in *Information Security for South Africa -2011*, 2011, pp. 1–9.

[39] J. S. Sengar and R. Sharma, "Review : Ad-Hoc Cloud Architecture & Modern Cryptography," *Int. J. Grid Distrib. Comput.*, vol. 9, no. 6, pp. 45–50, 2016.

[40] H. Rajaei and J. Wappelhorst, "Clouds & Grids: A Network and Simulation Perspective," in *Conference: 2011 Spring Simulation Multi-conference, SpringSim '11, Boston, MA, USA*, 2011, pp. 143–150.

[41] C. O. Rolim, F. L. Koch, C. B. Westphall, J. Werner, A. Fracalossi, and G. S. Salvador, "A Cloud Computing Solution for Patient ' s Data Collection in Health Care Institutions," no. ii, pp. 95–99, 2010.

[42] J. S. Sengar, "SURVEY : Reputation and Trust Management in VANETs," *Int. J. Grid Distrib. Comput.*, vol. 8, no. 4, pp. 301–306, 2015.

[43] S. Bouchenak, G. Gheorghe, G. Chockler, H. Chockler, and A. Shraer, "Verifying Cloud Services: Present and Future," *ACM SIGOPS Oper. Syst. Rev.*, vol. 27, no. 2, pp. 6–19, 2013.

[44] N. Sasikaladevi, "Trust Based Cloud Service Composition Framework," *Int. J. Grid Distrib. Comput.*, vol. 9, no. 1, pp. 99–104, 2016.

[45] V. Saranya, S. Ramya, R. Kumar, and T. Nalini, "Efficient and Parallel Data Processing and Resource Allocation in the Cloud by u sing Nephele ' s Data Processing Framework," *Int. J. Grid Distrib. Comput.*, vol. 9, no. 3, pp. 33–40, 2016.

[46] M. Irfan, M. Usman, Y. Zhuang, and S. Fong, "A Critical Review of Security Threats in Cloud Computing," in *2015 3rd International Symposium on Computational and Business Intelligence (ISCBI)*, 2015, pp. 105–111.

[47] P. Senthil, N. Boopal, and R. Vanathi, "Improving the Security of Cloud Computing using Trusted Computing Technology," *Int. J. Mod. Eng. Res.*, vol. 2, no. 1, pp. 320–325, 2012.

[48] F. S. Al-anzi, A. A. Salman, and N. K. Jacob, "New Proposed Robust, Scalable and Secure Network Cloud Computing Storage Architecture," no. May, pp. 347–353, 2014.

[49] V. Akshaya and T. Purusothaman, "Business Intelligence as a Service in Analysis of Academic Courses," *Int. J. Appl. Eng. Res.*, vol. 11, no. 4, pp. 2458–2467, 2016.

[50] S. Kai, T. Shigemoto, T. Kito, S. Takemoto, and T. Kaji, "Development of Qualification of Security Status Suitable for Cloud Computing System," in *Proceedings of the 4th international workshop on Security measurements and metrics - MetriSec '12*, 2012, p. 17.

[51] M. M. A. Ghosh, R. R. Atallah, and S. S. A. Naser, "Secure Mobile Cloud Computing for Sensitive Data: Teacher Services for Palestinian Higher Education Institutions," *Int. J. Grid Distrib. Comput.*, vol. 9, no. 2, pp. 17–22, 2016.

[52] K.-F. Ho, H. Hirai, Y.-H. Kuo, H. Meng, and K. Tsoi, "Indoor Air Monitoring Platform and Personal Health Reporting System: Big Data Analytics for Public Health Research," in *2015 IEEE International Congress on Big Data*, 2015, no. 2, pp. 309–312.

[53] A. Priya, T. Meena, and M. Devi, "Efficient Approach for Data Retrievability on Cloud Storage Systems," *IJSRSET*, vol. 2, no. 2, pp. 408–412, 2016.

[54] H. Suo, Z. Liu, J. Wan, and K. Zhou, "Security and Privacy in Mobile Cloud Computing," in *2013 9th International Wireless Communications and Mobile Computing Conference (IWCMC)*, 2013, pp. 655–659.

[55] R. Kumar and S. Rajalakshmi, "Mobile Cloud Computing: Standard Approach to Protecting and Securing of Mobile Cloud Ecosystems," in *Proceedings - 2013 International Conference on Computer Sciences and Applications, CSA 2013*, 2013, pp. 663–669.

[56] P. Srivastava, "Multiple Key Based Architecture to Secure Cloud Database," vol. 4, no. September, pp. 1–7, 2015.

[57] N. Ahmed and A. Abraham, "Modeling Security Risk Factors in a Cloud Computing Environment," *J. Inf. Assur. Secur.*, vol. 8, no. 2013, pp. 279–289, 2013.

[58] S. Mazur, E. Blasch, Y. Chen, and V. Skormin, "Mitigating Cloud Computing security risks using a self-monitoring defensive scheme," *Aerosp. Electron. Conf. (NAECON), Proc. 2011 IEEE Natl.*, pp. 39–45, 2011.

[59] P. Rohmeyer and T. Ben-zvi, "Managing Cloud Computing Risks in Financial Services Institutions," in *2015 Proceedings of PICMET '15: Management of the Technology Age*, 2015, pp. 519–526.

[60] K. Hashizume, D. Rosado, E. Fernández-Medina, and E. Fernandez, "An analysis of security issues for cloud computing," *J. Internet Serv. Appl.*, vol. 4, no. 5, pp. 1–13, 2013.

[61] D. G. Rosado, R. Gómez, D. Mellado, and E. Fernández-Medina, "Security Analysis in the Migration to

[62] Kashyap Rajesh and Sarika Sharma, "Security Challenges and Issues in Cloud Computing – The Way Ahead," *Int. J. Innov. Res. Adv. Eng.*, vol. 2, no. 9, pp. 32–35, 2015.

[63] B. S. Al-Attab and H. S. Fadewar, "Security Issues and Challenges in Cloud Computing," *Int. J. Emerg. Sci. Eng.*, vol. 2, no. 7, pp. 22–26, 2014.

[64] H. T. Dinh, C. Lee, D. Niyato, and P. Wang, "A Survey of Mobile Cloud Computing : Architecture , Applications , and Approaches," *Wirel. Commun. Mob. Comput.* -, no. Cc, pp. 1–38, 2013.

[65] M. Vermaat, S. Sebok, S. Freund, J. Campbell, and M. Frydenberg, *Discovering Computers 2016: Tools, Apps, Devices, and the Impact of Technology*. 2016.

[66] M. I. M. Hanifah, R. C. Omar, N. H. N. Khalid, A. Ismail, I. S. Mustapha, I. N. Z. Baharuddin, R. Roslan, and W. M. Z. Zalam, "Integrated Geo Hazard Management System in Cloud Computing Technology," *IOP Conf. Ser. Mater. Sci. Eng.*, vol. 160, p. 12081, 2016.

[67] A. A. Soofi and M. I. Khan, "Encryption Techniques for Cloud Data Confidentiality," *Int. J. Grid Distrib. Comput.*, vol. 7, no. 4, pp. 11–20, 2014.

[68] D. Tse, "Challenges on Privacy and Reliability in Cloud Computing Security," in *International Conference on Information Science, Electronics and Electrical Engineering (ISEEE), 2014*, 2014, pp. 1181–1187.

[69] L. Xu and J. Li, "Building Efficient Resource Management Systems in the Cloud : Opportunities and Challenges," *Int. J. Grid Distrib. Comput.*, vol. 9, no. 3, pp. 157–172, 2016.

[70] B. Al-shargabi and O. Sabri, "A study of Adopting Cloud Computing from Enterprise Perspective using Delone and Mclean IS Success Model," *Int. J. Comput. Sci. Inf. Secur.*, vol. 14 S1, no. February, p. 5500, 2016.

[71] A. Khrisna and Harlili, "Risk Management Framework with COBIT 5 and Risk Management Framework for Cloud Computing Integration," in *2014 International Conference of Advanced Informatics: Concept, Theory and Application (ICAICTA) Risk*, 2014, pp. 103–108.

Performance Evaluation of Web Application Server based on Request Bit per Second and Transfer Rate Parameters

Hafiza Samad[1], Hanizan S.H.[2], *Roshidi Din[3], Roslinda Murad[1], Asiah Tahir[1]

[1]School of Computing and Technological Science, Kolej Universiti Poly-Tech, 56100 Cheras, Kuala Lumpur
[2]Department of Information Technology, Kulliyyah Muamalat, Kolej Universiti INSANIAH, 09300 Kuala Ketil, Kedah
[3]School of Computing, College Arts and Sciences, Universiti Utara Malaysia, 06010 UUM Sintok, Kedah

*Corresponding Author's Email: roshidi@uum.edu.my

Abstract: An investigation found that the increasing popularity of information services distributed as the World Wide Web has resulted in three of scale that are charging excessive server due to the popularity of documents, loss bandwidth of the network due to transfer documents redundant. Therefore, this study is measure the performance of the Web application that embed with caching server in order to discover the influencing the it server toward speed and performance of the Web application

***Keywords*:** Web Application, Web Server, Web caching performance

1. Introduction

Nowadays the virtual world that became more important and the total number of Internet users in the world limit increase is in recent years. Online banking, ticketing, commerce and business are all over the Internet. Because of this, Web services have become important than ever. Web servers provide services that must maintain reliability and maintain tolerable in terms of server response time to customers in two orders to avoid customer frustration. It expected for developers have to stress test the server before it is available online.

This paper analyzed the web application performance based on load testing using performance testing tool that provides stress testing servers' response time in relation captured during request and response transactions using SDLC. Then, in order to validate whether the function is required, a free web load test called LoadUIWeb is used as benchmark. The project met and compared the results generated from both tools.

The problem in current scenario required the component based performance testing tool. Performance testing of components of a web application aims to evaluate the response time, utilization, and throughputs, as a function of the system description and workload parameters. The existing load and performance tools a mainly used to check the performance of CPU load for the entire web application. There are no such tools to confirm the total time taken to load the complete application and also individual components. The individual components may include the images and he files. All these issues results in the need of an efficient web based load testing tools for testing web application with individual components performance.

There are different providers of web servers in the market; some of them are free, mean while other is commercial. In this study, it will only be tested open source web servers. As said before, it will be chosen web servers that are highly customizable by the user and with enough interesting features.

This study applied to measure the performance of web application performance based on used the caching or without caching in order to discovered the comparison among the web application performance

2. Related Work

Generally there are numerous study linked to the evaluation and performance analysis of web application using different benchmarks. In this thesis, 10 related works have been discussed in order to reach the ideas of web application performance analysis. Their overall researches are used as guidelines to create additional experiment in terms of parameter experiment and additional tools used in load testing environment.

Zhu et al. [1] analysed the types, indicators and testing methods of the performance testing of the web, and then put forward some testing process and methods to optimize the strategy. Krizanic et al. [2] analyzed and compared several existing tools which facilitate load testing and performance monitoring, in order to find the most appropriate tools by criteria such as ease of use, supported features, and license. Selected tools were put in action in the real environment, through several web applications. For the case study one of the web applications is used in order to introduce different capabilities of tools, including distributed testing, assembling of test scenario from previously recorded usage scenarios, security support, results analysis, monitoring of key parameters, and reporting and alerting about changes of these parameters.

Andreolini et al. [3] introduced a simple queuing model to analyze the performance metrics of web server under varying traffic loads. This assists web server managers to manage their clusters and understand the trade-off between QoS and cost. In this proposed model two thresholds are used to control the scaling process. A discrete-event simulation (DES) is presented and validated via an analytical solution [4].

Bai and Hei [5] presented a performance evaluation method for mirror architecture of Web servers with a load balance mechanism. The performance evaluation method is based both on the usual queuing model for Web servers and on the corresponding quantitative analysis. By using the formula as presented in this paper, the author provided a procedure to evaluate the system response time for the mirror architecture of Web servers. In addition, they also used QNAT (Queue Network Analysis Tool) to do the simulation in order to verify the models for mirror Web servers and also provided the performance measurements of an experimental design of the Web servers. In conclusion from the analysis from simulation and measurements, they learned that the mirror architecture of Web servers can improve the service performance of Web requests.

Wang et al. [6] is proposed a usage model to simulate users' behaviors realistically in load testing of web applications, and another relevant workload model is proposed to help generate realistic load for load testing. The authors also demonstrated an eclipse-based load testing tool "Load Testing Automation Framework (LTAF)" which is based on these two models and can perform load testing of web applications easily and automatically. Furthermore, these models and tools were successfully applied into a representative web-based system from a big Corporation.

3. Methodology Framework

There are consist five stages involved to reach that shown in Figure 1 as following.

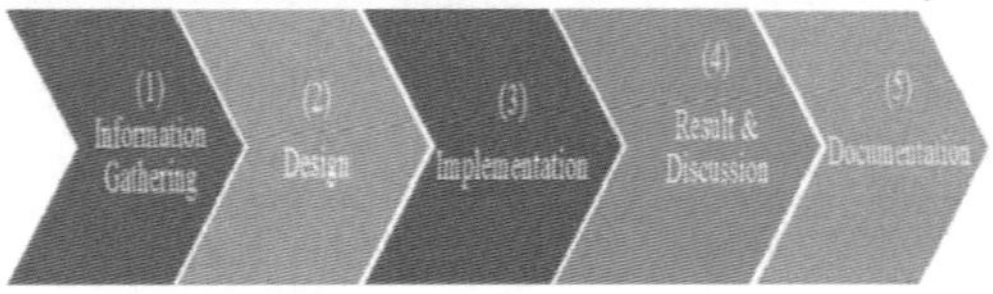

Figure 1: Methodology Framework

3.1 Information Gathering

This is the stage where all needed information is collected and analyzed. The information collected was related to configuration performance analysis tools in term of infrastructure and how the tools work. The outcome of this stage shall create a better understanding of the configuration performance analysis concept. In this stage, detailed study on related and previous work is conducted. From this stage, it also identified problems or limitations on configuration performance monitoring web application.

3.2 Design Stage

The design plan defines the steps to be taken to develop the test bed install performance monitoring tools. Details of the proposed design will be explained in the next sub section. The simplest example of using the cache module is single instance of the module. If the client sends a request, and a response for such request is cached, then the response will be sent from the module. This will result in reduced traffic inside the service and shorter response times. Figure 2 show the test bed architecture for this study.

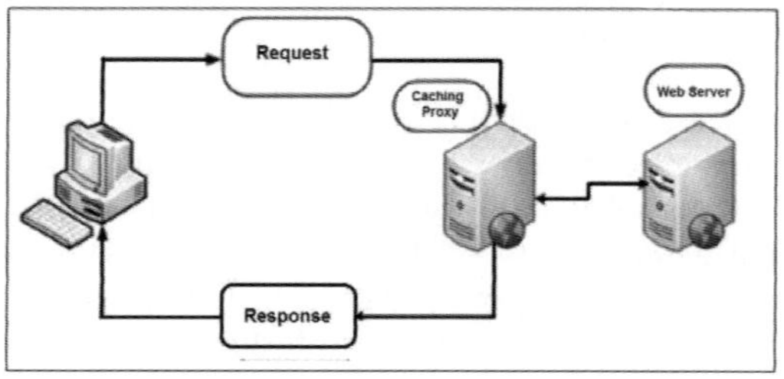

Figure 2: Test bed architecture design

3.3 Implementation Stage

In this stage, Webserver Stress Tool and LoadUIWeb are chosen as a performance analysis tools for deployment in this test bed environment. Suitable web server infrastructure is also produced to suit the organization's environment. The plan will include preparing the virtual machine and defining the architecture for web server. Then the implementation setup will take place based on the design produced.

3.4 Result and Discussion Stage

This stage shows deliverable in Table 1 as following.

Table 1. Result and Discussion

Task	*Activities*	*Deliverable*
• Producing project result and analysis	• To analyze and discuss from the tested results • Analyze the result graph by Comparison • Verify the result • Conclude the analysis of result.	• Complete project analysis

3.5 Result and Discussion Stage

This stage shows about documentation in Table 2 as following.

Table 2. Documentation

Task	*Activities*	*Deliverable*
• Documenting overall project activities, result and analysis.	• Producing project documentation consisting of an Introduction, Literature Review, Research Methodology, Implementation and Result.	• A complete project documentation

4. Implementation Web Application

This section elaborates the configuration and implementation the experiments starting from setting up the web server using Apache server.

4.1 Test bed deployment overview

Deployment of Web Server on a single node for the following purposes:
- Hosting simple web or CGI applications
- Developing and testing web applications

The following flowchart provides the schematic representation of how to deploy Web Server on a single node:

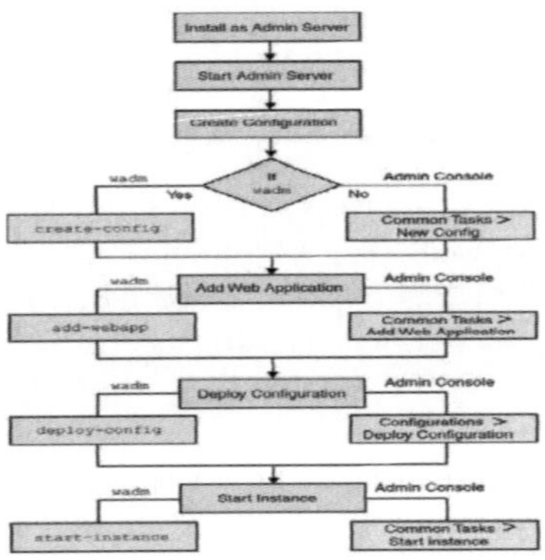

Figure 3: Flowchart representing the deployment of web server

Based on Figure 3 is showed the implementation that begin process installation admin server and create configuration of web server. The step of create configuration is based on web server deployment set up.

4.2 Web Server Configuration

In the preceding figure, the Web Server deployment set up comprises the following components:
- **Administration Server**- Administration Server is a specially configured web server instance. You can deploy web applications on the administration server.
- **Administration Node**- Administration Node is deployed on a node or a server/host within a server farm and has the ability to communicate with the remote Administration Server.
- **Configuration**- A configuration refers to a set of all configurable elements of a Web Server instance, such as web applications, configuration files, and search collection indexes. A configuration can be created, modified, or deleted.

- **Config-store-** This is the file system-based repository where all the configurations are stored.
- **Instance** - An instance refers to the environment of a web server on a given node, including its configuration, log files, and other runtime artifacts such as lock databases, caches, and temporary files.

5. Web Application Performance Results

In this section, comparison on web application performance is done using Apache Bench tool.

5.1 Request bit per second

In request bit per second it consist of four handle request, those are two handle 50 request with 5 and 10 request running that shown in Table 3 and 4 as following.

Table 3. Result of Request bit per second (-n 50, -c 5)

Average Transfer rate (ms) (-n 50, -c 5)	
With Caching	*Without Caching*
4.714	***1.706***

Table 4. Result of Request bit per second (-n 50, -c10)

Average Transfer rate (ms) (-n 50, -c 10)	
With Caching	*Without Caching*
1.892	***1.806***

Table 3 and 4 showed average request per second in Edu zone Web Application with different caching configuration. In Table 3 showed averaged requested per second increased with caching from 1.706 to 4.717, while in table 4 illustrated request per second also little bit increased from 1.806 to 1.892. Then, for two handle 100 requests with 5 and 10 requests running is shown in Table 5 and 6 as following.

Table 5. Result of Request bit per second (-n 100, -c 5)

Average Transfer rate (ms) (-n 50, -c 5)	
With Caching	*Without Caching*
4.50	***1.71***

Table 6. Result of Request bit per second (-n 100, -c10)

Average Transfer rate (ms) (-n 50, -c 10)	
With Caching	*Without Caching*
6.202	***2.308***

Based on Table 5 and 6 is showed that handle 100 requests per second also increased with caching where with caching in 5 second running increased from 1.71 until 4.50, while with 10 seconds running increased 6.202. The comparison the average handle 50 and 100 showed in Figure 4.

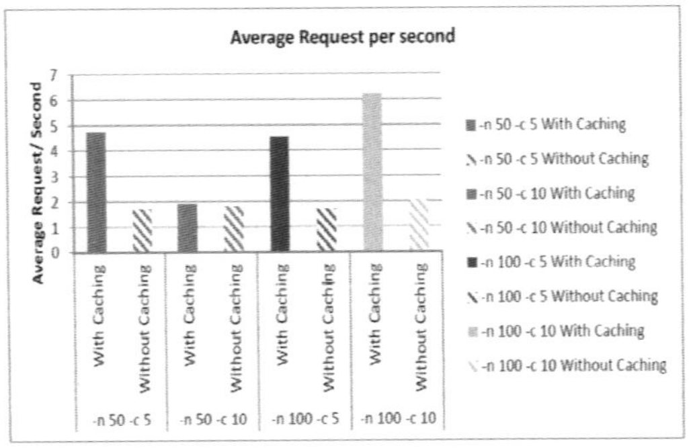

Figure 4: The average request bit per second between number of request and requests running concurrently

Based on Figure 4, it showed the comparison of average request per second between different parameter value which are number of request and requests running concurrently. It is proved that web caching has improved 23% performance of Eduzone web application if we increase the concurrency level from 5 to 10 or so (which would otherwise starve the server completely).

5.2 Transfer rate

It has created two tests to measure the transfer rate for Eduzone web application with different parameter value those are handle 50 request, with 5 request running and handle 100 request with 10 requests running that shown in Table 7 and Table 8.

a) Handle 50 request, with maximum of 5 request running concurrently

Table 7. Result of Transfer Rate (-n 100, -c 10)

Average Transfer rate (ms) (-n 100, -c 10)	
With Caching	*Without Cachin g*
72.572	***25.77***

Table 7 showed the transfer rate recorded for Eduzone Web Application at. The experiment found out that with caching technique had improved 64.5% of the of web application performance (increased server transfer rate from 25.744 to 72.572.)

b) Handle 100 request, with maximum of 10 request running concurrently

Table 8. Result of Transfer Rate (-n 100, -c 10)

Average Transfer rate (ms) (-n 100, -c 10)	
With Caching	*Without Caching*
105.271	***30.814***

Table 8 showed the average transfer rate for Eduzone web application is very similar to the first test above even though it uses a different parameter values. As a conclusion, web caching had improved the performance of web application (increased server transfer rate from 30.814 to 105.27). As the comparison result transfer rate is shown in Figure as following.

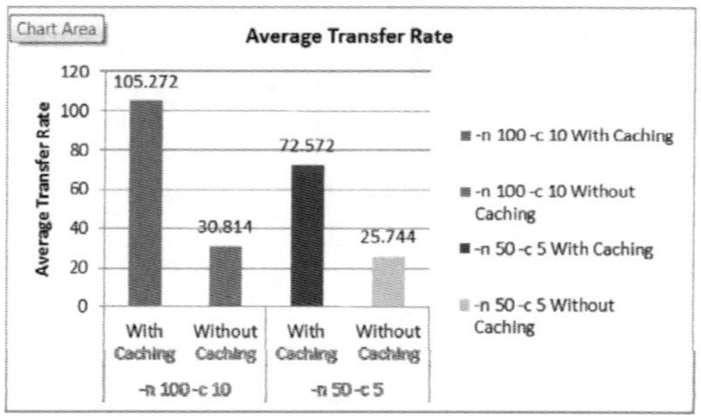

Figure 5. Comparison Graph of average transfer rate between number of request and requests running concurrently

In Figure 5 showed the comparison of average transfer rate between different parameter value which are number of request and requests running concurrently. It is proved that web caching has improved 31.05% performance of Eduzone web application .Then, if it is increased the concurrency level from 5 to 10 or so (which would otherwise starve the server completely).

From the conceptual study, caching realized a performance was increased the transfers of data repeatedly, while a caching system might realize a performance increase upon the initial (typically write) transfer of a data item. The performance increase is due to buffering occurring within the caching system so buffering itself sometimes increases transfer performance. Moreover, Web caching is one of the key strategies that had been proved to increase the web application performance.

6. Conclusions

This research into web page performance quantifies what web page attributes effect response times the most. The most important factors in speeding up web page response times are minimizing the number of embedded objects. Those are clearly discovering was Caching able to improve response times in request bit per second and also transfer rate.

References

[1] P. Yungming, X. Mingna, Load Testing for Web Applications. 2009.
[2] J. Krizanic, A. Grguric, M. Mosmondor, P.Lazarevki. Lord Testting and Performnce monitoring tools in use with AjAX based web applications.
[3] M. Androelini, M. Colajanni, P. Velente. Design and Testing of Scalable Web-based Systms with performance constraints. 2005.
[4] P. Chung, S. H. Chung, C. Hui. A Web Server Design Using Search Engine Optimization Techniques for Web Intelegence for Small Organizations. 2012.
[5] W. Bai, H. Wei. Model Based Load Testing of Web Applications. 2010
[6] X. Wang, B. Zhou, Li W. Model Based Load Testing of Web Applications. 2010.

Risk and Threat via Online Social Network among Academia at Higher Education

Hanizan Shaker Hussain, Roshidi Din, Nik Zulkarnaen Khidzir, Khairul Azhar Mat Daud, Suzastri Ahmad

[1]School of Computing, College Arts and Sciences, Universiti Utara Malaysia, 06010 UUM Sintok, Kedah

*Corresponding Author's Email: roshidi@uum.edu.my

Abstract. The evolution of information and communication technologies (ICT) nowadays has changed the life style of human living. The current modern societies have adopted ICT as an important thing that they are really needed in their life, especially as a tool to be used for communications activity. However, unfortunately ICT also exposed its user in circumstances of risk, threats and vulnerability. This paper will discuss the risk and threats to the users who are using social media as a medium to communicate. In this paper, the fraction of user will be divided by two types which are gender and working experience. The data that obtained from the distributed of questionnaires among respondent will be analysed by using SPSS. Data will be analysed by using two-way ANOVA statistic in order to examine the significant level in between gender and working experience as an independent variable in this study with the level of threats in cybersecurity risk towards lecturers who are working in higher education institutions in Malaysia. This article also will provide an empirical data and will be referred to another researcher in the future for their further research perhaps.

1. Introduction

Cybersecurity risk means the risk which has occurred to the users who are using information and communication technologies in several of aspect. In this paper, discussion will be focused on the lecturers who are working in higher education institution in Malaysia as the user for ICT technologies. In nature, lecturers deal with ICT technologies in teaching activities such as e-learning, research activity, searching for varied types of information through internet, communication in research activities by applying video conferencing to discuss with their colleagues who are located at various places, send and deliver document through email and a variety of activities regarding to their nature of work. By exposing themselves through online networking via varieties of online medium, this will expose lecturers to the threats. However, many lecturers or users are really didn't realize with the risk that they will face. This paper will show the empirical outcome regarding to the level of threats in cybersecurity risk among lecturers at higher education institution in Malaysia.

2. The integration of Cybersecurity in Malaysian's Higher Education

The method of face-to-face learning needs to be reviewed in order to enhance the effectiveness of instructional regarding to the growing up of tremendous information and communication technology in this age [1]. Learning in 21st century is about implementing digital technologies [1]. Social media is a very important medium that must be mastered by the lecturers who are working in higher education institution. By using social media, those who are working as lecturers will tremendously build up their reputation in the academic world. By applying social media, lecturers will build their reputation and bridging in career opportunities such as to market themselves in academicians' activity such as research, publications, co-operation, collaboration and so on [2]. By using social media, scholars will position themselves in popularity of ranking among academicians around the world. In learning and teaching activity for example, the current methods of learning such as problem based learning, inquiry learning and so on which is in group of constructivism learning will be implemented more effectively by the integration of all kind of social medias which is known as e-learning in order to deliver and convey knowledge among students in higher education institutions.

Unfortunately, Social Digital Media could lead to several critical cybersecurity risks that could cause serious impact to the lecturers itself in context of threats among lecturers in higher education institution in Malaysia [2]. Therefore, the extensive awareness programme on cybersecurity risk should be conducted continuously in higher learning institutions in order to minimize the possibility of the risk happened and mitigate their impact more effectively. The entire higher learning institutions eco-sphere will get benefit from this programme as a preventive approach to manage the confidentiality, integrity and availability of the information transported through cyberspace.

3. Cybersecurity Challenges and Requirements

Online social digital media has become very useful equipment in the era of digital nowadays. By the used of online social media in a right way and proper technic and well strategies, the learning activities will be delivered more effectively and it's concurrent with the needs of educations around the globe. However, Nik Zulkarnaen at.al [2] asserted that besides the advantages functions and capability offered by this technology, it's still have several issues and challenge that need to be contemplated by other scholars.

Hanna, et. al. [3] asserted that, users will be able to create online communities to share information, ideas, personal messages, and other content through the social media platform. The rapid evolution of social media was build up and create the ecosphere of communication regardless of border via a variety of social media platforms that currently exist [4]. The emerging of digital technology to promote social media as a communication platform had given many effects to the societies' ways of life. In education aspect, a traditional or conventional way of learning and teaching need to be reviewed and do a little bit of adjustment in order to ensure our education system is relevance and parallel with other countries that adopt e-learning as their main agenda in their country. The challenging to begin the implementation of social media as a main platform in education activities in higher education institution is to break the conventional practices among all education communities. By the way, after through with a long challenging process to shift the perception and prejudice among all members in educations communities and with a proper strategic plan proposed by the ministry of education, more than 15 years ago, now Malaysia is one of the countries in Asia that have some advanced technologies which is able to support ICT and provide internet and social media to the citizens.

Generally, social media considered as part of the internet society ecosystem that serve an effective communication channel for cyber community. Besides the unbelievable opportunities offered through this technology, cybersecurity risks considered as emerging challenges to look into seriously [5]. Regarding to that issue and the expected risk of cybersecurity, this paper will discuss about the factors of threats that may be faced by the lectures who are working at higher education institution in Malaysia. By looking and examine the level of threats in cybersecurity risk among lecturers at higher education institution in Malaysia, hopefully the finding of this study will give a bit of view to the authorities who are making policies and strategic planning in order to prevent unpleasant matters from occurs to our nation in the future.

Cybersecurity risk is currently becoming serious issues in digital social media due to the increasing number of social media population growth [2]. Valerică Oana [6], asserted that the spectrum of the risks are really wide and unpredictable. Social media allows social engineer use the psychological manipulation of people into performing actions confidential information for the purpose of information gathering, fraud or system access [7]. Digital Social Media becomes the source of information for Social Engineer to capture and harvest the useful information for the purpose of the cyber-attack [8]. Online Social Media might highly contribute to cybersecurity risks on Critical Information Infrastructure and need to be identified and measure in order to understand the level of severity for each of them so that will be managed effectively [2]. The capability of Online Social Media act as communication platform among cyber community and its could also become a phishing pond for cyberwarfare and cybercrimes. For extreme cases, the impact of the cyberwarfare and cybercrime might cause human personal security, safety and psychological impact in nature [9].

4. Research methodology

This research was done by using quantitative methods. Data obtained through the administration of questionnaire to hundred (100) respondents. The respondents were comprised of lecturers from Higher Education Institution in Malaysia. Respondents were randomly selected. The instrument was developed after several constructs involving the formation of threats in cybersecurity risk among lecturers who are using online social media as a communications medium to work. The value of validity and reliability of the instruments were measured through the pilot test. A pilot test was conducted on 30 lecturers from a selected higher education institution in Malaysia separately (Table 1). Through the pilot study, the obtained values of reliability and validity of the instrument was very high. The value of Cronbach Alpha, α obtained was 0.962. This study was conducted in the state of Selangor. The respondents comprised of 100 lecturers from higher education institution in Malaysia. A brief demographic data distribution of respondents involved in this study is presented in table 2 below.

Table 1. Reliability Test Result: Cronbach's Alpha Value

Cybersecurity Risk Factors	Items	Cronbach's Alpha Value	Total Num. of Respondents (N)
Threats	18	0.962	100

Table 2. Lecturers' participation by gender

Gender	Frequency	Percent
Male	39	39%
Female	61	61%
Total (N)	100	100%

Table 2 show the gender tabulation among respondents who are involved in this study. Respondents consisted of 52 (17.3%) males and 248 (82.7%) females.

Table 3. Working Experience percentage

Working Experience	Frequency	Percent
5 years and below	10	10%
6 years to 10 years	30	30%
11 years to 14 years	23	23%
15 years to 20 years	19	19%
21 years and above	18	18%
Total (N)	100	100%

Table 3 shows the distribution of fractional group of respondents who have different periods of working experience. There are five (5) fraction of working experience among respondents, first is lecturers who had been teaching in the last five (5) years, second is lecturers who had been working in the last six (6) years to ten (10) years, third is lecturers who have working experience in the last eleven (11) years to fourteen (14) years, fourth is the lecturers who have working experience in fifteen (15) years to twenty (20) years and lastly is the lecturers who are working more than twenty-one (21) years.

Data obtained from the questionnaire related on threats in cybersecurity risk to the lecturers who are working at higher education in Malaysia was analyze by using two-way analysis of variance (ANOVA). This analysis was made by using SPSS statistical package for social science (SPSS Version 21). ANOVA analysis was conducted to determine the joint effect of two independent variables on one dependent variable.

Table 4. Descriptive Statistics
Dependent Variable: Risk Threats of Cyber Security at Higher Education in Malaysia

Teachers' Gender	Teaching Experiences	Mean	Std. Deviation	N
Male	5 years and below	2.7889	.39752	5
	6 years to 10 years	3.0556	.72344	15
	11 years to 14 years	3.3765	.49906	9
	15 years to 20 years	2.5278	.66782	2
	20 years and above	3.3194	1.13146	8
	Total	3.1225	.75361	39
Female	5 years and below	3.1111	.31427	5
	6 years to 10 years	3.2815	.87878	15
	11 years to 14 years	3.1270	.58892	14
	15 years to 20 years	3.3529	.69443	17
	20 years and above	3.3056	.69401	10
	Total	3.2559	.68676	61
Total	5 years and below	2.9500	.37811	10
	6 years to 10 years	3.1685	.79917	30
	11 years to 14 years	3.2246	.55769	23
	15 years to 20 years	3.2661	.72188	19
	20 years and above	3.3117	.88441	18
	Total	3.2039	.71282	100

Table 5. Two-way ANOVA test results

Source	Sum of Squares	df	Mean Square	F	Sig. (p)	Partial Eta Squared
Gender	.821	1	.821	1.569	.214	.017
Working Experiences	1.367	4	.342	.653	.627	.028
Gender * Working Experiences	1.904	4	.476	.909	.462	.039
Error	47.125	90	.524			
Total	1076.793	100				

* Value is significant when p <0.01
a. R Squared = .063 (Adjusted R Squared = -.002)

According to Table 5, it was found that there was no significant difference in level of threats to cybersecurity risk in higher education of Malaysia among lecturers based on gender (F (1, 90) = 1.57, p = 0.21), in which the effect size is very small (eta squared = 0.02). This means that the level of threats in cybersecurity risk is the same between of genders. Comparison of the threats on cybersecurity risk in higher education in Malaysia toward working experience, found that no significant differences in the level of threats on cybersecurity risk in higher education of Malaysia based on working experience (F (1, 90) = 0.653, p = 0.63), in which the effect size is very small (eta squared = 0.03). In the perspective of higher learning institution, personal working experience does not influence much the cybersecurity risk threats level.

According to Table 5, there was no significant interaction effect between gender and working experience on the threats of cybersecurity risk in higher education of Malaysia (F (1, 90) = 0.909, p = 0.039). This means that the level of threats of cybersecurity risk in higher education of Malaysia is not dependent on the join effect gender and working experience. Therefore, further analysis is not needed in order to describe the lecturers' threats of cybersecurity risk based on gender and working experiences.

5. Conclusion

The current trend of social communications is supported by digital technologies. Every kind of communication activities among members of society was applying the social media as an effective and practical platform. In the scope of educations, e-learning is an indicator to show the priorities of social media in learning and teaching activity. Regarding to that, almost all lecturers at higher education institutions in Malaysia have been using social media as an important and main tool of their teaching activity. In other words, lecturers were exposed with the global ecosphere according to their academicians' activities such as research, collaboration, publications, and conferences, e-learning and so on.

The studies finding discovered the level of threats among lecturers in view of gender and working experiences factors among the lecturers who are working at higher education institution in Malaysia. The finding shows that the level of threat toward lecturers is not really significant with the gender and working experience. Regarding to the finding of this research, the authorities who are intended to do strategic planning or proposed some kind of policies regarding to the cybersecurity risk doesn't need to worry to thinking about the factors of gender and working experience. The two independent variables that are highlighted in these studies were being concluded that they do not give any effect to contribute threats in cybersecurity risk.

References

[1] Khairul Azhar Mat Daud 2016 A New Interaction Design Model For Digital Chronological Infographic Development *Elixir Edu. Tech.* 90 (2016) 37522-37527 ISSN: 2229-712X

[2] Nik Zulkarnaen Khidzir, Ahmad Rasdan Ismail, Khairul Azhar Mat Daud, Mohamad Shahfik Affendi Abdul Ghani, Suriatini Ismail and Asrul Hery Ibrahim. Human 2016 Factor of Online Social Media Cybersecurity Risk Impact on Critical National Information Infrastructure *7th International Conference on Applied Human Factors and Ergonomics* (AHFE 2016)

[3] Hanna, R, Rohm, A, and Crittenden, V L 2011 We're all connected: The power of the social media ecosystem *Business Horizons* **54**(3) 265–273.

[4] Colliander, J, and Dahlén, M 2011 following the fashionable friend: The power of social media. *Journal of Advertising Research* **51**(1) 313–320.

[5] National Cyber Security Policy 2013 *An Assessment. Institute for Defence Studies and Analyses* http://www.idsa.in/idsacomments/NationalCyberSecurityPolicy2013_stomar_2608 13

[6] Valerică GREAVU-ŞERBAN, Oana ŞERBAN 2014 Social Engineering a General Approach *Informatica Economica Journal* **18** no. 2 2014.

[7] Anderson and Ross J 2008 *Security engineering: a guide to building dependable distributed systems (2nd ed.)* Indianapolis IN: Wiley 1040. ISBN 978-0-470-06852-6. Chapter 2, 17.

[8] Kaplan, A. M, and Haenlein, M. 2010 Users of the world, unite! The challenges and opportunities of Social Media. *Business Horizons* **53**(1) 59–68.

[9] Tang, Q, Gu, B and Whinston, A B 2012 Content contribution for revenue sharing and reputation in social media: A dynamic structural model *Journal of Management Information Systems* **29**(2) 41–76.

Preserve Imperceptibility and Robustness Performance on Steganography Technique based on StegaSVM-Shifted LBS Model

Hanizan Shaker Hussain [1], Roshidi Din[2], Rusdi Idrus[2]

[1]Department of Information Technology, Kulliyyah Muamalat, Kolej Universiti INSANIAH, 09300 Kuala Ketil, Kedah

[2]School of Computing, UUM College of Arts and Sciences, Universiti Utara Malaysia, 06010 Sintok, Kedah, Malaysia

Email: drhanizan@insaniah.edu.my, roshidi@uum.edu.my, rushdi@uum.edu.my

Abstract. In image steganography, the most popular and widely used techniques is the least significant bit (LSB) that hide data into a cover-image in a spatial and discrete cosine transform (DCT) domain as well. Beside the LSB technique, there is other technique that is also influential i.e support vector machine (SVM) normally used to strengthen the embedding algorithm. Whatever techniques used in the image steganography field,the main purpose is to keep the existence of the secret-message secret. This paper designing the new model is proposed called StegaSVM-Shifted LSB model in DCT domain to preserve the imperceptibility and increase the robustness of stego-images. The StegaSVM-Shifted LSB model that has been proposed that utilize HVS and embedding technique through Shifted LSB showed a good performance.

Keywords: Image steganography, DCT model, SVM-Shifted

1. Introduction

The importance of information security has increased due to the increased use of computers and the Internet. This increase of utilization also include contributions by one of its exciting subfields i.e. information hiding. Information hiding is a technology where the secret-messages are hidden inside other files in the form of an image [1][2][3]. There are two popular important research fields namely steganography and watermarking. Steganography is a process that involves hiding a message in an appropriate carrier file for example in an image or in an audio file [4] whereas watermarking is used to place a hidden message (e.g. copyright information) in digital media, like images, music or software [5]. This paper focuses on steganography as the domain implementation area. Then, medium of steganography that used in this study is image medium for the process development of the technique in steganography.

Image steganography is a field in information hiding, besides watermarking, that hides and secures secret-messages written inside an innocuous-looking cover-image file. As a sub-field in information security, it is currently used to secure transmission and information storage in many fields like in the military and medical fields. The generic image steganographic model, as shown in Figure 1, contains two main processes, embedding and extracting.

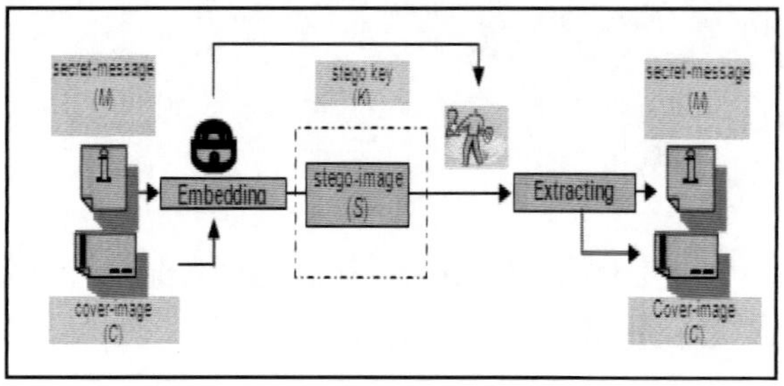

Figure 1. Generic image steganographic model

Based ond Figure 1, A secret-message, M will be embedded in a cover-image, C; the result is a stego-image, S and it is normally sent through a public channel together with a stego-key, K. Then, the recipient will extract the secret-message from the stego-image by using the same stego-key. The terminologies, *secret-message*, *cover-image*, *stego-image* and *stego-key* were agreed to be used in steganoraphy research area at the First International Workshop on Information Hiding held in Cambridge, UK.

In designing steganography algorithms, three requirements are generally considered namely imperceptibility, robustness and payload capacity [4][6][7]. Nevertheless, in image steganography, the most important requirements are imperceptibility and robustness [6][8][9]. These two requirements are important to resist especially two kinds of statistical attacks that is passive and active steganalysis. Many researchers concentrates on passive steganalysis,the attacks only takes actions when the stego-image is found suspicious [10][11]. Meanwhile, active steganalysis is the attacks that attempt to foil all possible covert communications [12][13]. Therefore, this study used StegaSVM-Shifted LSB model is introduced which takes into account. It design consist two main models, StegaSVM classification and StegaSVM-Shifted LSB. In StegaSVM classification, the classified cover-image will be utilized and the non-smooth area, the most appropriate and imperceptible is preferred to be used during the embedding process. StegaSVM classification also will be applied in the extracting process in order to extract the right secret message.

Meanwhile, with the StegaSVM-Shifted LSB model, the proposed technique called Shifted LSB technique will ensure only the least significant bits of DCT coefficients of a cover-image will be utilized to exploit the human visual system (HVS). This Shifted LSB technique is applicable only when it is a case where the original embedding locations a smooth area. Then the technique is applied to shift the embedding position to the nearest best neighbors' LSB. The strength of Shifted LSB technique will be utilized not only in the embedding but also in extracting process. Figure 2 shows the model which highlights the relationship between the two main processes involved and the requirements (i.e. robustness and imperceptibility) to be achieved.

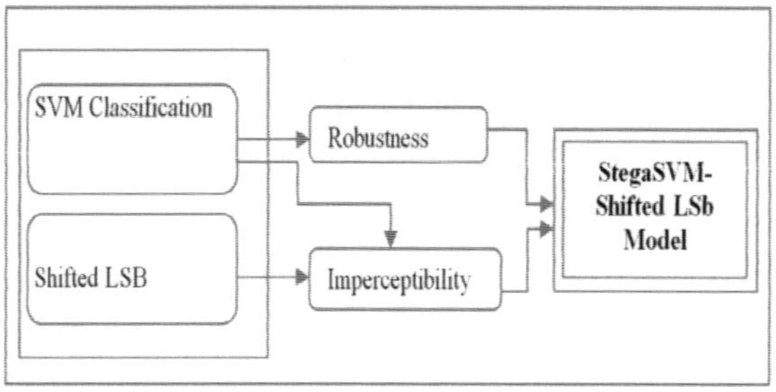

Figure 2. Basic idea of StegaSVM-Shifted LSB model

2. Support Vector Machines

The theory of SVM is based on the idea of structural risk mimization (SRM). SRM is an inductive principle for model selection used for learning from finite training data sets. Basically, SVM uses a linear separating hyperplane to create a classifier with a maximal margin [14]. In cases where given classes cannot be linearly separated in the original input space, the SVM first transforms the original input space into a higher dimensional feature space [15]. In Figure 3 is shown a non-linearly hyperplane separating. This transformation can be achieved by using various non-linear mapping such as polynomial, radial basis function and sigmoid. Then, the task of a SVM in finding the optimal separating hyperplane in this feature space is simply done [15].

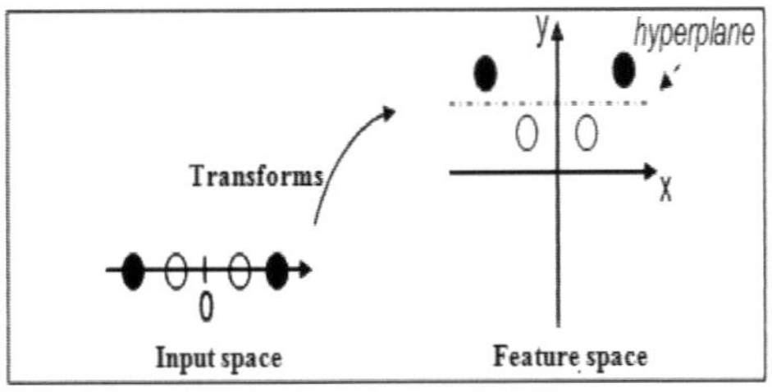

Figure 3. Feature Mapping Transformation

All SVMs algorithms have been applied in watermarking and no work has been done in steganography. These types of algorithms are normally used in the embedding and extracting procedures. All works concentrate on imperceptibility to utilize the SVMs good ability in learning the relationship between hidden-information and cover-image that is useful to be used in embedding and extracting functions [16][18]. All algorithms use images as a secret-messages and applied secret key for robustness and security. All algorithms have used localization approach in their process because the changes are normally done on less sensitive channels to the human eye (the blue and luminance channels) [16][17]. Then the average pixel intensity (localized modification) based on the certain shapes will be computed to determine the changes on the channels. All works use RBF kernel to map the training samples to a higher dimensional feature space.

SVM has a proper learning and generalization capabilities and it shows to be very resistant to the over-fitting problem [19]. Unfortunately, the SVM is rather sensitive to outliers or noises since the penalty term of SVM treats every data point equally in the training process. SVM was originally introduced by Vapnik within the area of statistical utilization [15]. In statistical learning theory, the SVM has been developed for data classification and prediction and it is used in wide range applications such as character recognition [20], text categorization [21], face detection in images [22] and information hiding in images [23].

A thorough review of the literature has revealed that all SVM works in information hiding have been done to utilize the SVM good learning ability to memorize the relationship between a random coefficient and its neighboring coefficients [24][25]. It is mainly used to classify the cover-image and find the suitable area for embedding. Even though SVM has lately been used extensively for image classification in watermarking [25][26] but none has been done in steganography. Only Wu et al. [27] has attempted to introduce SVM in steganography but his work is merely to support vector regression (SVR) to predict the pixel values for embedding rather than classifying the cover-image.

SVM was first initiated by Yu et al. [28] in watermarking by introducing the SVM-based Color Image Watermarking (SCIW) that utilizes the image feature i.e. pixel intensity which is calculated from three difference types of shapes of neighboring pixels for classification. His attempt is fruitful in utilizing SVM to classify the cover-image into smooth and non-smooth areas by using 1024 bits as training bits. For security purposes, he employs a pseudo-random number generator (PRNG) to randomly pick the embedding position before the secret-bits is embedded by modifying the blue channel of color components. The trained SVM is exploited again in the extracting process to acquire the right secret-message. Then, there have been further research to utilize the ability of SVM in spatial domain as proposed by Yu et al. [28].

3. Model of Design

The proposed design begin with a general model that is called StegaSVM-Shifted LSB that involves two main models StegaSVM classification and StegaSVM-Shifted LSB. Hopefully, through this general model description will give the reader a rough idea about the proposed design model. Then, a more detailed description of each model StegaSVM classification and StegaSVM-Shifted LSB can be followed after that. But for the StegaSVM-Shifted LSB, its two main sub-models will be explained one by one starting with StegaSVM-Shifted LSB embedding then StegaSVM-Shifted LSB extracting for the sake of simplicity.

3.1 StegaSVM-Shifted LSB Model

The proposed StegaSVM-Shifted LSB model is as shown in Figure 4. This model contains all main processes involved combining StegaSVM classification, Shifted LSB embedding and Shifted LSB extracting.

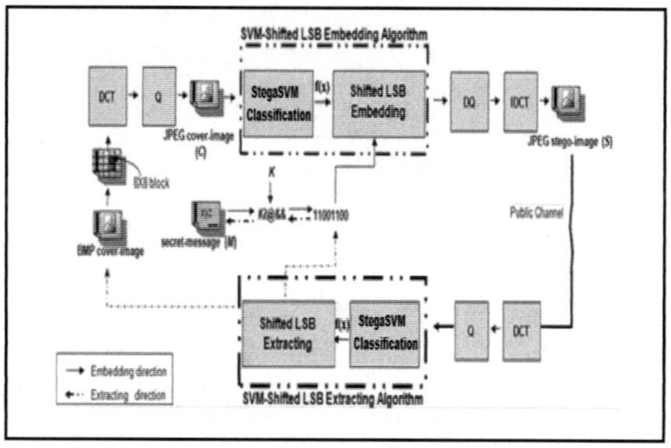

Figure 4. Process of StegaSVM-Shifted LSB Model

Based on Figure 4 this model begin with the BMP cover-image, *C* is first divided into 8x8 non-overlapping blocks. Then, it transformed from space domain to transform domain in order to make it more robust to attacks and remain imperceptible to the human eyes. In the quantization step, all DCT coefficients are divided by the predefined quantization values to modulate the different spectral components on the image. This permutation can equalize the spread of insignificant coefficients. The hybrid techniques of the StegaSVM classification and Shifted LSB will be done through the StegaSVM-Shifted LSB embedding algorithm for cover-image classification and secret-message embedding respectively. The secret bits from the transformed secret-message, *M* will be embedded in the medium frequency of the insignificant coefficients based on the trained SVM before it will be permutated and generated by using a user-defined key, *K*.. The output, stego-image (*S*) will go through inverse DCT and quantization before it can be sent via public channel. The extracting process at the recipient side will involvea reverse of the embedding process called StegaSVM-Shifted LSB extracting algorithm as it employs a symmetric model.

3.2 StegaSVM-Shifted LSB Classification

The StegaSVM classification model works when 256x256 JPEG grayscale cover-image is being read. Image features such as luminance, edge and entropy values will be extracted from the cover-image. Then a dataset will be constructed by classifying it into SVM format. The dataset is comprised of the following set:

SVM::Dataset(label= {1,-1} Data ={luminance, edge,entropy})

The basic concept of StegaSVM classification model is shown in Figure 5 as follow.

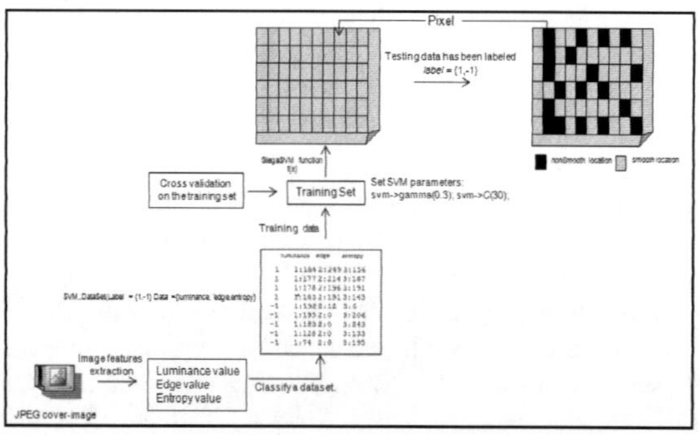

Figure 5. Basic concept StegaSVM-Shifted LSB classification model

The SVM classification contains with two class labels of 1 (smooth) and -1 (non-smooth). While the data is made up of three attributes, namely luminance, edge and entropy arranged in specific columns in Figure 6. The first column for SVM labels, the second column to the first attribute, the second attribute of the third column and the fourth column for the third attribute.

	luminance	edge	entropy
1	1:184	2:249	3:156
1	1:177	2:214	3:167
1	1:178	2:196	3:191
1	1:183	2:191	3:143
-1	1:192	2:12	3:6
-1	1:195	2:0	3:206
-1	1:183	2:0	3:243
-1	1:128	2:0	3:133
-1	1:74	2:0	3:195

Figure 6. SVM dataset format

The available dataset can then be used as training data in the training set. At this time, SVM parameters should be identified and set i.e. gamma γ and penalty parameter C (e.g. gamma = 0.3 while C = 30). In addition, the percentage of training data points also need to be determined. Not to forget that a cross validation approach should also be done here in order to get an SVM model that is efficient and suitable for image steganography. Through this approach, various combinations of parameters of SVM will be used and tested. The best SVM parameters pairs will be chosen to be used in the testing set.After that, as a result of cross validation activity, a StegaSVM function $f(x)$ will be generated. Then $f(x)$ will be utilized on testing data in the embedding process which eventually resulted a stego-image as a result of the exploitation of smooth (-1) and non-smooth (1) areas.

In designing a suitable StegaSVM classification model with the field of image steganography is being implemented. Designing an appropriate model with the field that is to be investigated (i.e. image steganography) is crucial in order to achieve the desired goals (i.e. a highly imperceptible and robust image steganographic model).

3.3 StegaSVM-Shifted LSB Embedding Model

The StegaSVM-Shifted LSB embedding model process is shown in Figure 7 in this paper.

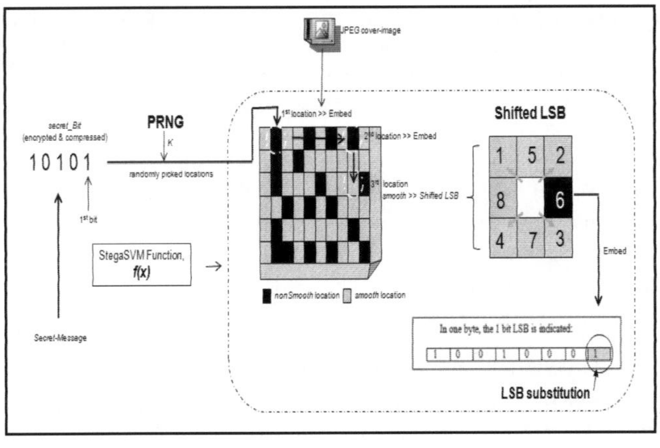

Figure 7. StegaSVM-Shifted LSB embedding model

Based on Figure 7 the sub-model, a a secret-message is first converted into a sequence of bits which is then encrypted and compressed. In Figure 8, sequence of these bits is marked as *secret_Bit*. Then, by using pseudo-random number generator (PRNG) with a secret key, *K* for any secret-bit, *secret_Bit* = {0,1}, a random embedding location will be obtained. According to Yu *et al.* [28], PRNG is used in the research especially in information hiding because of its simplicity and security purposes. The

IOP Conf. Series: Journal of Physics: Conf. Series **1018** (2018) 012009 doi:10.1088/1742-6596/1018/1/012009

embedding process is really depends on the type of embedding location, whether it is smooth (*smooth location*) or non-smooth (*nonSmooth location*). If it is *nonSmooth location* then embed *secret_Bit* by applying the LSB substitution. If it's otherwise, do the Shifted LSB technique that is, either go to 8-neighbors' LSB following the specific sequence, if found any of the first the *nonSmooth location*, embed *secret_Bit*.

4. Discussion

In this model, various types of cover-images (i.e different image complexities) have been used; Lena, Baboon, Uitm and Clock. As for the secret-message, the size that has been used is 1024bits for all types of cover-images. Meanwhile, for the cover-image classification, the value of C has been set to 30 while the value of γ is set at 0.5. Then, in Table 1 shows cover-images with high image complexities, Lena dan Baboon have recorded the highest PSNR; 49.86 and 49.33 respectively. This is followed by Uitm cover-image with PSNR of 48.89 whereas the cover-image that has low image complexities, that is Clock, recorded the lowest PSNR with 47.68.

Table 1. PSNR of different types of cover images

Types of cover-images	Different Image Complexities	PSNR
Lena	High	49.86
Baboon	High	49.33
Uitm	Moderate	48.89
Clock	Low	47.68

High image complexities actually represents image with more striking colors, more shades, the combination of lots of light and dark colors, or there will be many lines or edges. For example, Lena and Baboon images have many different color combinations that cover almost the entire surface of the image. So, from here it has formed more edges and line areas that can be called as non-smooth areas that are more secured to use for embedding in the Shifted LSB operation. So with that, they have a higher PSNR values compared to other images of more than 49dB. Meanwhile,Uitm cover-image has a moderate level of image complexities because it has a uniform color of black only.It is similar to the Clock cover-image that has plain bright colors, that is why the smooth areas are higher compared than non-smooth area. So, when there are not many non-smooth areas that can be utilized in the Shifted LSB operation, then the recorded result of PSNR value is quite low for both cover-images.

This display the StegaSVM-Shifted LSB model, all cover-images, even the one with different cover-image complexities (high or low image complexities) are still able to maintainthe quality of a cover-image after the embedding process. This can be done through the capability of the proposed techniques of StegaSVM classification to successfully classify the cover-image into a better location for embedding. In addition, through the Shifted LSB operation,using a better location that is identified for embedding and extracting process was proven a successful effort. Embedding technique randomly and uniformly throughout the cover-image applied by the proposed model also enhance cover-image quality. In addition, this model has also been successful in reducing the impact of non-random changes to avoid secret-message from being easily detected. The increase in secret-message size can lead to quality decline in certain cover-image. It becomes clearer when the size of the secret-message is equal to 2048bits; it shows that the cover-image with the size of 256x256 could not afford to accommodate the large secret-message.

5. Conclusions

A model called StegaSVM-Shifted LSB was designed and developed in this chapter.In this chapter, the design was initiated with a description of the main model, namely StegaSVM-Shifted LSB. Then, the next description is broken down into three sub-models that are StegaSVM classification, StegaSVM-Shifted LSB embedding and StegaSVM-Shifted LSB extracting. The StegaSVM classsification model had the activities involved were to extract image features, construct training dataset, select kernel, implement cross-validation and execute SVM training set. The technique would shift the embedding position to the nearest best neighbors' LSB when it wasa case where the location was a smooth area. In the StegaSVM-Shifted LSB extracting model, the same Shifted LSB technique would also be applied in finding the actual embedding location for extracting process.

References

[1] J. He, S. Tan, T. Wu. On the Security of Steganographic Techniques. International Congress on Image and Signal Processing, (2008) 716-719.

[2] M. T. Siponen, H. Oinas-Kukkonen, H. A Review of Information Security Issues and Respective Research Contributions. Database for Advances in Information Systems, (2007) 60-80.

[3] C.-C. Chang, T.-S. Chen, L.-Z. Chung. A Steganographic Method Based Upon JPEG and Quantization Table Modification. International Journal of Information Sciences (2002) 123-138.

[4] A. Cheddad, J. Condell, K. Curran, P. M. Kevitt, P. M. Review: Digital Image Steganography: Survey and Analysis of Current Methods. Journal Signal Processing, (2010). 727-752.

[5] Y.-G Fu, R.-M. Shen, L.-P. Shen, X.-S. Lei. Reliable Information Hiding Based on Support Vector Machine. International Journal of Informatica, (2005) 333-346.

[6] N. Sathisha. Embedding Information in DCT Coefficients Based on Average Covariance. International Journal of Engineering Science and Technology, (2011) 3184-3194.

[7] P. M. Santi, M.K. Kundu. Genetic Algorithms for Optimality of Data Hiding in Digital Images. Bio-Inspired Information Hiding, (2009) 361-373.

[8] M. A. Younes, A. Jantan, A. A New Steganography Approach for Image Encryption Exchange by Using the Least Significant Bit Insertion. International Journal of Computer Science and Network Security (2008) 247-254.

[9] Y. K. Lee, L. H. Chen. (2000). A Secure Robust Image Steganographic Model. The Conference on Information Security, (2000) 275-284.

[10] L. Bin, H. Junhui, H. Jiwu, Q. S. Yun. A Survey on Image Steganography and Steganalysis. Journal of Information Hiding and Multimedia Signal Processing , (2011) 142-172.

[11] J. He, S. Tan, T. Wu. On the Security of Steganographic Techniques. International Congress on Image and Signal Processing, (2008) 716-719.

[12] R. Din, A. Shamsudin. Digital Steganalysis: Computational Intelligence Approach. International Journal of Computers, . (2009) 161-166.

[13] R. Chandramouli. A Mathematical Framework for Active Steganalysis. *Multimedia Systems*, (2003) 303–311.

[14] T. Joachims. Support Vector and Kernel Methods. In C. S. Department, SIGIR 2003 Tutorial. Cornell University (2003).

[15] V. Kecman. Learning and Soft Computing. London: The MIT Press. (2001).

[16] S. Huang, W. Zhang. Digital Watermarking Based on Neural Network and Image Features. *International Conference on Information and Computing Science*, (2009) 238 – 240.

[17] M. Jamzad, Z. Z. Kermani. Secure Steganography Using Gabor Filter and Neural Networks. *Transactions on Data Hiding and Multimedia Security*, (2008) 65-71.

[18] K. Naoe, Y. Takefuji. Damageless Information Hiding Using NN on YcbCr Domain. *International Journal of Computer Science and Network Security, (2008).*

[19] C. Campbell. Kernel Methods: A Survey of Current Techniques. *New Computings* (2002) 63-84.

[20] C.-L. Liu, K, Nakashima, H. Sako, H. Fujisawa. Handwritten Digit Recognition: Benchmarking of State-of-the-Art Techniques. *Pattern Recognition*, (2003) 2271-2285.

[21] T. Joachims. Text Categorization with Support Vector Machines: Learning with Many Relevant Features. *European Conference on Machine Learning*,(1998) 137-142. Chemnitz, Germany.

[22] E. Osuna, R. Freund, F. Girosit. Training Support Vector Machines: An Application to Face Detection . *IEEE Computer Society Conference on Computer Vision and Pattern Recognition,* (2002) 130-136. San Juan , Puerto Rico: IEEE Computer Society.

[23] L. Li, W.-Y Ding, J.-Y. Li. A Novel Robustness Image Watermarking Scheme Based on Fuzzy Support Vector Machine. *The 3rd IEEE International Conference on Computer Science and Information Technology* (2010) 533 – 537.

[24] S. Ramly, S. A. Aljunid, H.S. Hussain, H. S. SVM-SS Watermarking Model for Medical Images. *Digital Enterprise and Information Systems*, (2011) 372-386.

[25] F. Meng, H. Peng, Z. Pei, J. Wang. A Novel Blind Image Watermarking Scheme Based on

Support
Vector Machine in DCT Domain. *International Conference on Computational Intelligence and Security* , (2008) 16 - 20.

[26] H.-H. Tsai, H.-C. Tseng, Y.-S. Lai. Robust Lossless Image Watermarking Based on A-Trimmed Mean Algorithm and Support Vector Machine. *The Journal of Systems and Software*, (2010) 1015–1028.

[27] H.-C. Wu, K.-C. Liu, J.-D. Chang, C.-H. & Huang. An Image Steganographic Scheme Based on Support Vector Regression (SVR). *Proceedings of the Eighth International Conference on Intelligent Systems Design and Applications,* (2008) 519 - 524.

[28] P.-T. Yu, H.-H. Tsai, D.-W. Sun. Digital Watermarking of Color Images Using Support Vector Machine. *National Computer Symposium (NCS'03), (2003).* Taichung.

IOP Conf. Series: Journal of Physics: Conf. Series **1018** (2018) 012011 doi:10.1088/1742-6596/1018/1/012011

VisualUrText: A Text Analytics Tool for Unstructured Textual Data

Zuraini Zainol[1], Mohd T.H. Jaymes[2] and Puteri N.E. Nohuddin[3]

[1]Computer Science Department, Faculty of Science and Defence Technology, Universiti Pertahanan Nasional Malaysia, Kem Sungai Besi, Kuala Lumpur, Malaysia
[2,3]Institute of Visual Informatics, Universiti Kebangsaan Malaysia, Bangi, Malaysia

zuraini@upnm.edu.my, tarhamizwan123@outlook.com, puteri.ivi@ukm.edu.my

Abstract. The growing amount of unstructured text over Internet is tremendous. Text repositories come from Web 2.0, business intelligence and social networking applications. It is also believed that 80-90% of future growth data is available in the form of unstructured text databases that may potentially contain interesting patterns and trends. Text Mining is well known technique for discovering interesting patterns and trends which are non-trivial knowledge from massive unstructured text data. Text Mining covers multidisciplinary fields involving information retrieval (IR), text analysis, natural language processing (NLP), data mining, machine learning statistics and computational linguistics. This paper discusses the development of text analytics tool that is proficient in extracting, processing, analyzing the unstructured text data and visualizing cleaned text data into multiple forms such as Document Term Matrix (DTM), Frequency Graph, Network Analysis Graph, Word Cloud and Dendogram. This tool, VisualUrText, is developed to assist students and researchers for extracting interesting patterns and trends in document analyses.

1. Introduction

In the era of big data, government, corporates and business and even society are contributing to produce and use substantial amount of data. Data is often referred to facts, number or text, which can be processed by computer [1]. Data is limitless and presented everywhere in the universe. In computer system, data is referred to input symbols or signals that are stored and processed by the computer to produce information. Recently, the advancement of computing technology and devices has led to the rapid growth of data sources such as social networking applications, images, online news, email messages, medical records, etc. This huge amount of data is believed to have interesting patterns and valuable hidden knowledge within this data. This situation has attracted a large number of researchers to study on data mining techniques to extract interesting information and knowledge which can assist the stakeholders in decision making and prediction particularly in business, risk management, and medical research, etc. Data mining is an analytics process that is designed to explore large amount of data in search of consistent patterns or systematic relationships between data features and then validate the findings by applying the detected patterns to new subsets of data [2]. Technically, data mining is a process of finding correlations or patterns among dozens of fields in a large number of data. Data mining covers wide areas of research in text mining, web mining, image mining, medical mining, graph mining, etc.

The increment of text data has also increased its interest in our daily life. According to [3], 80-90% of future world's data growth is expected from the unstructured text databases or document databases.

In other words, most of available information is kept as text databases, which contain a high commercial potential value in discovering and predicting interesting patterns and trends. When it comes unstructured text, the recurrent text processing always involves text analysis and content analysis. By using data mining tools for example RapidMiner, Weka, Sentinel Visualizer, R, Orange and many more, users are able to analyze data from different dimensions, categorize, and summarize the text features and relationships can be identified. Technically, data mining is a process of finding correlations or patterns among dozens of fields in a large number of data [4]. The aim of this paper is to develop a text analytics tool that is capable of extracting, processing, analyzing the unstructured text data and visualizing the clean text data into multiple forms such as Document Term Matrix, Frequency Graph, Network Analysis Graph, Word Cloud and Dendogram. This tool called VisualUrText is developed to assist the students and researchers in knowledge discovery of analyzing documents. The rest of this paper is organized as follows. Section 2 explores Knowledge Discovery in Databases and Text Mining. Section 3 discusses the framework for visualizing unstructured text data. Section 4 discusses the experiment and results. Finally, we conclude this paper with future work in Section 5.

2. Related Work and Background

In this section, we discuss several related topics on Knowledge Discovery in Databases (KDD), text mining and some reviews on text analyzer tools.

2.1. Knowledge Discovery in Databases

Knowledge discovery in databases (KDD) or sometimes also referred as data mining is a process of discovering interesting patterns (non-trivial, implicit, previously unknown and potentially useful) or knowledge from a large of databases [5]. The KDD involves several processes such as data preparation and selection, data cleansing, incorporating prior knowledge on data sets and interpreting accurate solutions from the observed results. As well as the prominence on gathering data growing around the world, there is a requirement for another era of strategies, different techniques and algorithms to help analysts, decision makers, researchers and managers in squeezing out the valuable patterns from increasing volumes of data. These tools and techniques are the subject of the appearing field of knowledge discovery in databases.

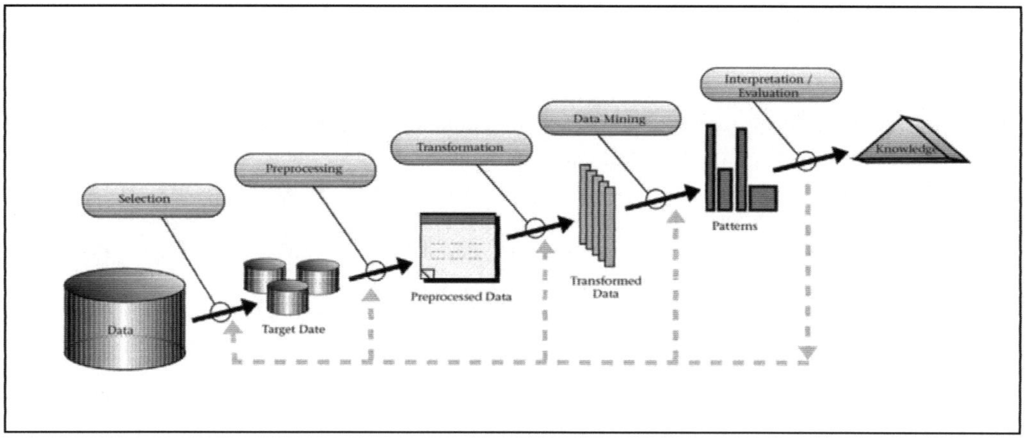

Figure 1. Overview of steps constituting the KDD process adopted from [6]

Figure 1 shows a general KDD process which consists of seven main essential steps which starts from raw data repository to new knowledge. The steps are described as follows [7]:

- Data Cleaning: This phase deals with missing values, noisy data and inconsistent data. For example, remove advertisement from web pages, detect and correct inaccurate records, normalize text converted from binary formats, deal with tables, etc.

- Data Integration: The phase involves with combining or merging data from multiple data stores.
- Data Selection: This is an important phase as it will select the relevant data to the analysis.
- Data Transformation: This phase exchanges the required data into a compatible format using data mining methods such as smoothing, aggregation, generalization, normalization and discretization, etc.
- Data Mining: This phase will apply intelligent methods to extract useful patterns and knowledge.
- Pattern Evaluation: Interesting patterns data represent the identified knowledge.
- Knowledge Representation: The discovered knowledge is then displayed to users in a visual form. Generally, visualization techniques will be applied to assist users in understanding and interpret information.

Although these are the main steps in any KDD process, some of the steps could be executed collectively during the actual process. For example, data cleaning, data transformation, data selection and data transformation can be combined together. Basically, the first four steps are listed above can be considered as data preparation for data mining process whereas the last three steps can be seen as post-processing step [8].

2.2. Text Mining[1]

Text Mining (TM) or text data mining refers to an analytic process, which is designed to explore the unstructured text documents in search of useful information and knowledge hidden from a large amount of text resources [9]. In business world, this useful information may reveal interesting patterns or semantic relationships and trends in large volumes of unstructured data. In TM, patterns are extracted from natural language text rather than databases which can lead to new business opportunities and improvements in processes. Previous studies in the field of knowledge data discovery have estimated that 80% of company available information is in the form of text documents [3]. Based on this fact, it is believed that TM has a higher potential value compared to data mining. TM is an extension of data mining, which involves multiple disciplines areas such as information retrieval (IR), Statistics, Web Mining, Computational Linguistics and Natural Language Processing (NLP) [10-11]. TM can also be described as intelligent text analysis, text data mining and knowledge discovery in text (KDT). TM is different compared to web search or directory search site. In the web search, people are usually looking for what has already been known or written by someone. In comparison, the aim of the TM is to discover something that is nobody knows or not yet written down. It mines information and knowledge from a mountain of texts [12]. TM has been widely applied in many real world applications and is reported to have been applied in large-scale problems in both data science and business [13]. For example:

- prevent cybercrime – using TM intelligence and anticrime applications would make the internet crime prevention easier. For example, [14] discovered social relations between criminals from unstructured text using text mining method. [15] proposed a method to analyze criminal networks in a set documents that is able to discover both direct and indirect relationships among members in a crime community.
- Sentiment analysis – text analytics has been utilized in analyzing large volumes of unstructured data such as extracting opinions, emotions, sentiments and their relations with products. For instance, [16] proposed a technique to group the forums into several clusters for online communities.
- biomedical applications – TM can be applied as knowledge based search engine doe medical text. For example, PubGene is an application of TM in biomedicine.
- Documents clustering and frequent patterns – TM is also capable of extracting frequent patterns and interesting relationships in set of documents. For example, [12] introduced a keyword processing technique to cluster hadith chapters as references for Islamic scholars. [9] investigated the frequent patterns and interesting relationships in hadith chapters to that can benefits Muslim scholars in their daily Islamic life and practice using the technique of Association Rule Mining. [17-18] proposed a framework to discover significant text patterns

exist in military peacekeeping documents.

3. The Framework For Visualizing Unstructured Text Data

Figure 2 illustrates the framework for Visualizing Unstructured Text Data. It consists of two main phases: (i) Unstructured Text Data Pre-processing and (ii) Analyzing and Visualizing Selection of Keyword.

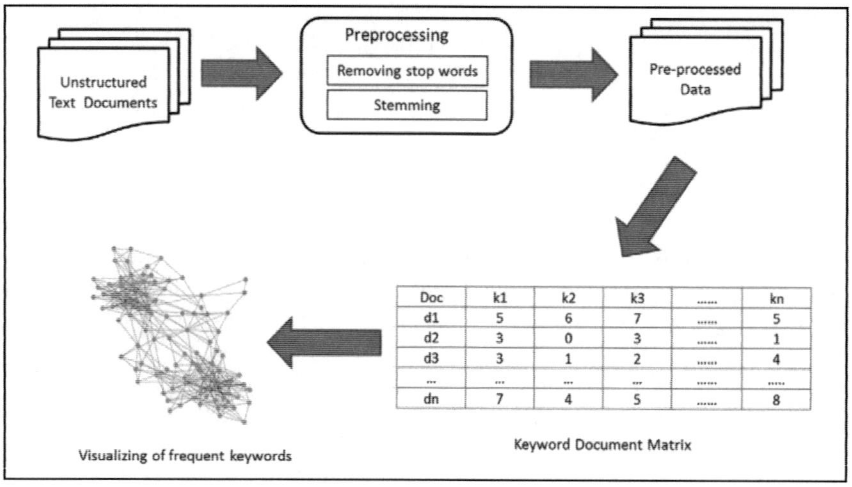

Figure 2. A Framework for Visualizing Unstructured Text Data

The first component consists of the pre-processing task while the second component is mainly focused on analyzing and visualizing the selection keywords/terms. The input materials being processed in this framework was a collection of text data. The details of each module are discussed in the following sub-sections.

3.1. Unstructured Text Data Pre-processing Phase

For investigation purposes, this experiment will apply a sample of six text documents (files) which are selected randomly from articles on data mining. Text pre-processing phase is the most important step in text analysis as it removes errors and inconsistencies from data in order to improve the quality of data. In this phase, the extracted documents need to be processed in order to facilitate the text analysis. Real world data is often incomplete and inconsistent. The pre-processing data can improve the accuracy and effectiveness of text mining [19]. In this study, the text pre-processing method consists of the following two basic steps: removing stop words and reducing words into their grammatically root or stem through stemming. Stop words is a list of words that normally does not carry any meaning and information such as 'a', 'about', 'after', 'all', 'before', 'such', 'the', 'some', 'she', etc. The most frequently used lists of stop word are available at www.ranks.nl/stopwords. These words need to be removed because they are most probably irrelevant for this study. The second step of pre-processing is stemming. A large document may consist of many words that often have similar common root. Stemming is a process that chops off the end of words without considering linguistics features of the words. In other words, stemming is a process of getting the basic word of a particular sentence. For instance, the words 'used', 'using', and 'uses' would all be stemmed to the same word that is 'use'. In this study, we apply one of the most popular algorithms for stemming English words, which is known as Porter algorithm. For example, *Data mining, the extraction of hidden predictive information from large databases, is a powerful new technology with great potential to help companies focus on the most important information in their data warehouses*. The example of steps of pre-processing is presented in Table 1:

Table 1. Steps of Preprocessing Unstructured Text Data

IOP Conf. Series: Journal of Physics: Conf. Series **1018** (2018) 012011 doi:10.1088/1742-6596/1018/1/012011

Function	R Scripts
Removing stop words whitespace	*data mining extraction hidden predictive information large databases powerful new technology great potential help companies focus most important information data warehouses*
Stemming	*data mine extract hid predict inform larg databas power new technolog great potenti help compani focu most import inform data warehouse*

3.2. Analyzing and Visualizing Selection of Keyword

In this phase, all cleaned text data is then ready for selecting keywords and visualizing them in multiple forms such as DTM, frequency graph, network analysis graph and dendogram. In keyword selection, important keywords are selected and their presence in each file is counted. DTM is a conceptual representation that represents the relationship between m keywords and n files. In this study, each row represents for a file whereas each column represents a keyword. Each cell contains values that show the number of occurrences of the keyword in the file. For instance, if the keyword appears in a file then the matrix entry for corresponding to that particular row and column is set as 1. If the keyword appears twice then it will be recorded as two in that particular matrix entry. Many data mining techniques can be carried out on DTM such as clustering, classification, and association analysis. The next step is to illustrate the top frequent terms/keywords visually using a bar plot. The number of minimum frequency can be set up by users by inserting the number of frequency at the bottom of this application tool. After building the DTM, the importance of terms/keywords with a word cloud can be easily produced. A word cloud, is a visual representation for text data, typically used to depict keyword metadata (tags) on websites, or to visualize free form text. Tags are usually single words, and the importance of each tag is shown with font size or colour.

4. Experimental Result And Analysis

This section discusses the experimental result for the proposed visual unstructured text data (VisualUrText). In this experiment, six text files were selected and uploaded into the VisualUrText tool for cleaning before they can be applied into DTM. As mentioned earlier in Section 3, the text pre-processing involves two main steps such as removing stop words and stemming.

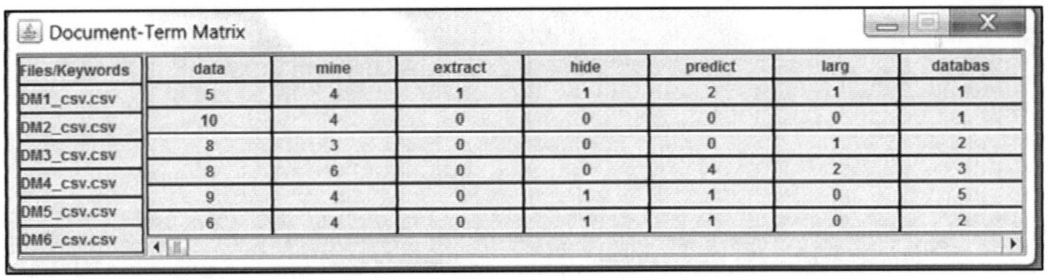

Document-Term Matrix

Files/Keywords	data	mine	extract	hide	predict	larg	databas
DM1_csv.csv	5	4	1	1	2	1	1
DM2_csv.csv	10	4	0	0	0	0	1
DM3_csv.csv	8	3	0	0	0	1	2
DM4_csv.csv	8	6	0	0	4	2	3
DM5_csv.csv	9	4	0	1	1	0	5
DM6_csv.csv	6	4	0	1	1	0	2

Figure 3. The screenshot for Document-Term Matrix (DTM)

In remove stop words phase, stop words such as articles ('a', 'an', 'the', etc.), conjunctions ('and', 'but', etc.), qualifiers ('yet', 'however', etc.) are removed whereas the stemming phase will reduce the a word into the root form such as the keyword 'mining' can be reduced into 'mine'. The next step is to generate a DTM. DTM is applied to represent the relationship between keywords and files, where each column refers to a keyword and each row represents files that were used in this experiment. The number of occurrences of each keyword in each file is presented in matrix to that particular row and column. As shown in Figure 3, the DTM is composed of 340 keywords and 6 files. The extract from DTM presents the keyword 'data' has the highest frequencies that occurs 55 times in 6 files while the keyword 'extract' presence once in file number 1 but not present in any of other 5 files. The complete list of keywords occurrence in six files is presented in Figure 3.

IOP Conf. Series: Journal of Physics: Conf. Series **1018** (2018) 012011 doi:10.1088/1742-6596/1018/1/012011

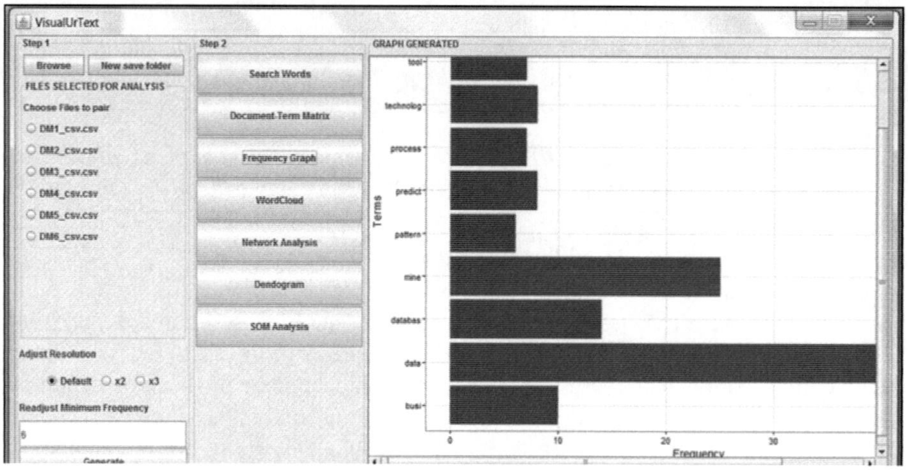

Figure 4. The screenshot for the occurrences of keywords using bar plot

The next step is to find the frequent term with frequency is not less than six. In order to present the most frequent words visually, a bar plot is applied in this study. From the TDM, we derive the frequency of terms. As seen in Figure 4, the two most frequent keywords are ‘data’ and ‘mine’ are clearly shown in the bar plot where the minimum number of frequency is set as six in this experiment.

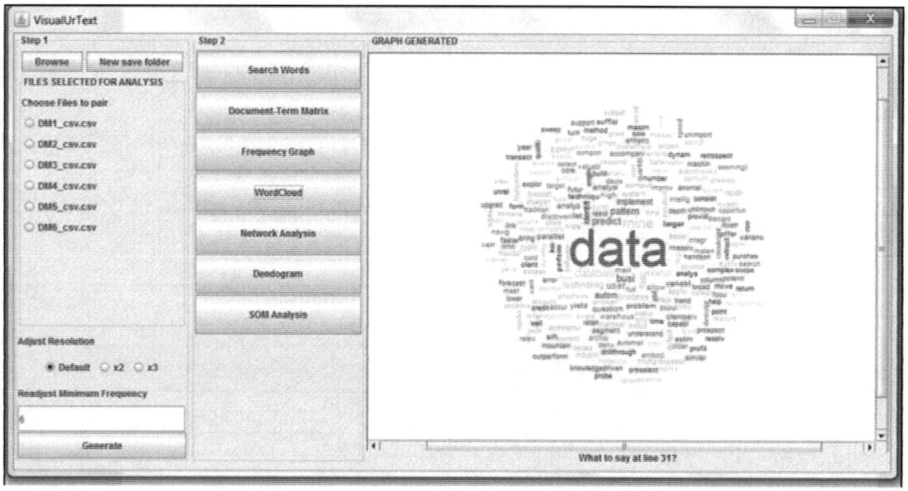

Figure 5. The screenshot for the occurrences of keywords using word cloud

After building a DTM, the importance of keywords with a word cloud can be displayed. Figure 5 illustrates that a word cloud (or tag cloud, text cloud) that highlight the most frequently used keywords in text data. Word cloud is easy to understand and readable as it provides multiple choices of colors that symbolizing the keywords with different sizes. As presented in Figure 5, the keywords ‘data’, ‘mine’, ‘databas’ are the top three most important keywords in six files, which validates that six files present information that related to data mining and databases. Some other important words are ‘busi’, ‘predict’, ‘technolog’ and ‘process’ which present the definition of data mining is technology that predict trends for business. What we can observe here is that the size of keywords is corresponding to the frequency of the keyword.

IOP Conf. Series: Journal of Physics: Conf. Series **1018** (2018) 012011 doi:10.1088/1742-6596/1018/1/012011

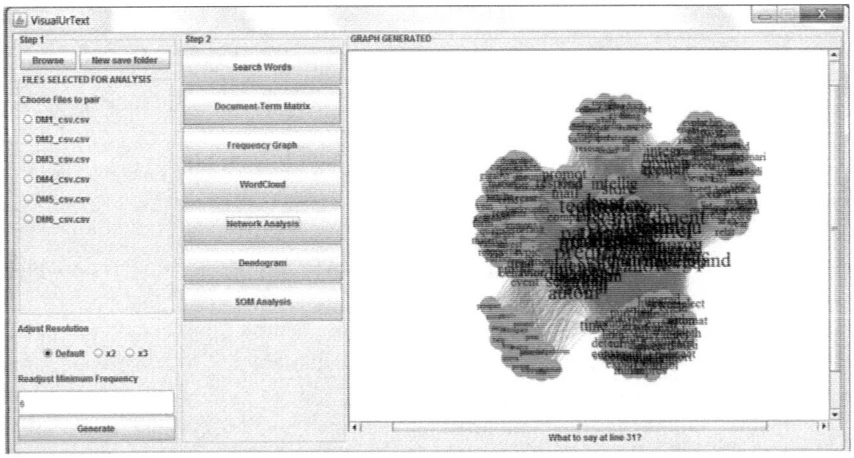

Figure 6. The screenshot for the occurrences of keywords using network analysis graph

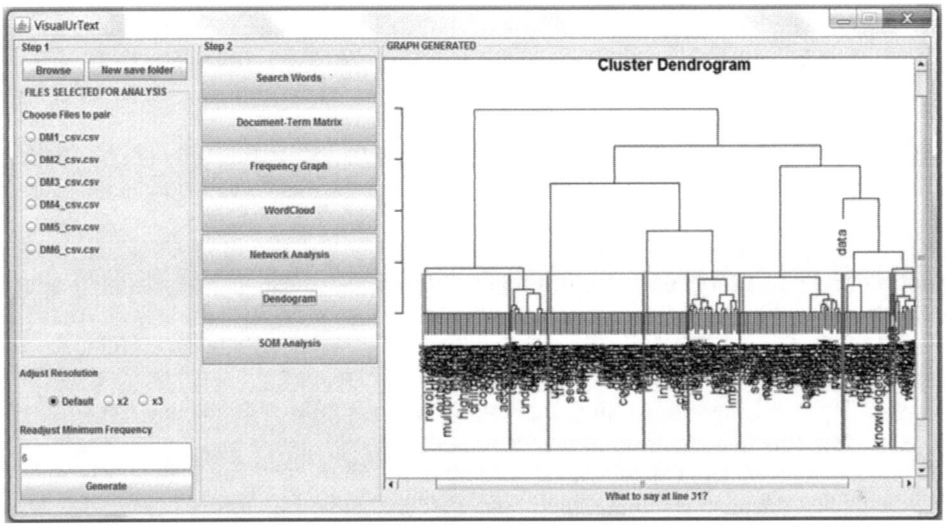

Figure 7. The screenshot for the clustering keywords

Figure 6 shows the network of keywords which is based on their co-occurrence in six files. The network analysis graph is plotted to show the relationships between frequent keywords. In order to make the graph more readable, colors and font sizes are set in this experiment. Figure 7 presents a dendogram (tree structured graph) that visualizes the result of hierarchical clustering analysis. It presents the arrangement of the clusters for 340 keywords in six files. From the dendrogram, we can see that the cluster analysis has eight clusters where the keyword 'data' is placed in the seventh group.

5. Conclusions And Future Work

This study proposes a text analytics tool that is capable of extracting, processing, analyzing the unstructured text data and visualizing the clean text data into multiple forms such as Term Document Matrix, Frequency Graph, Network Analysis Graph, Word Cloud and Dendogram. For future work, this research will extend the VisualUrText by incorporating the following features: (i) the lemmatization feature for reducing an inflectional or derivationally related word form to its base form, (ii) the enhancement of dendogram feature in term of customization of selected important keywords in documents before the visualization of result can be carried out. The setting of colours and font sizes would also assist the users in reading the graph and (iii) to upgrade the VisualUrText tool into web-based applications.

References

[1] Mourya, S. K., and Shalu Gupta. “Data mining and data warehousing”. Alpha Science International, Ltd, 2012.
[2] Punjabi, B., & Honale., S., “A Research on Big Data Analysis & Processing With Data Mining Techniques”, 2015
[3] EMC Education Services. “Data Science and Big Data Analytics: Discovering, Analyzing, Visualizing and Presenting Data”, John Wiley & Sons, USA, 2015
[4] “Data Mining: What is Data Mining?”. http://www.anderson.ucla.edu/faculty/jason.frand/teacher/technologies/palace/datamining.htm
[5] Osmar R. Zaiane. “Introduction to Data” Mining”. https://webdocs.cs.ualberta.ca/~zaiane/courses/cmput690/notes/Chapter1
[6] Fayyad, U., Piatetsky-Shapiro, G., & Smyth, P. From Data Mining to Knowledge Discovery in Databases. AI magazine, 1996, 17(3), 37.
[7] Han, J., Kamber, M. and Pei, J., “Data Mining: Concepts and Techniques”, (3rd ed.). Morgan Kaufmann Publishers Inc., San Francisco, CA, USA, 2011.
[8] Fivelstad, Ole Kristian. "Temporal text mining: the ttm testbench." (2007).
[9] Z. Zainol, P. N. E. Nohuddin, M. T. H. Jaymes, and S. Marzukhi, "Discovering “interesting” keyword patterns in Hadith chapter documents," in 2016 International Conference on Information and Communication Technology (ICICTM), 2016, pp. 104-108.
[10] Gupta V, Lehal G.S. A survey of text mining techniques and applications. Journal of emerging technologies in web intelligence. 2009 Jan 8;1(1):60-76
[11] Irfan R, King CK, Grages D, Ewen S, Khan SU, Madani SA, Kolodziej J, Wang L, Chen D, Rayes A, Tziritas N. “A survey on text mining in social networks”. The Knowledge Engineering Review. 2015 Mar 1;30(02):157-70.
[12] Nohuddin, P.N.E., Zainol, Z., Chao, K.F., Tarhamizwan, M., and Nordin, A. “ Keyword based Clustering Technique for Collections of Hadith Chapters.” International Journal on Islamic Applications in Computer Science And Technologies – IJASAT. 2015, 4(3), pp.11-18
[13] Fayyad, Usama, Gregory Piatetsky-Shapiro, and Padhraic Smyth. "From data mining to knowledge discovery in databases." AI magazine 17.3 (1996): 37.
[14] Elyezjy, N. T., & Elhaless, A. M. Investigating Crimes using Text Mining and Network Analysis. International Journal of Computer Applications, 2015,126(8).
[15] Al-Zaidy, R., Fung, B. C., Youssef, A. M., & Fortin, F. Mining criminal networks from unstructured text documents. Digital Investigation, 2012, 8(3), 147-160.
[16] Li, N., & Wu, D. D. Using text mining and sentiment analysis for online forums hotspot detection and forecast. Decision support systems, 2010, 48(2), 354-368.
[17] Zainol, Z., P.N.E. Nohuddin, T.A.T. Mohd and O Zakaria.Text analytics of unstructured textual data: a study on military peacekeeping document using R text mining package. International Conference on Computing and Informatics, 1–7. School of Computing, UUM. 2017
[18] Zainol Z., Marzukhi S., Nohuddin P.N.E., Noormaanshah W.M.U., Zakaria O. Document Clustering in Military Explicit Knowledge: A Study on Peacekeeping Documents. In: Badioze Zaman H. et al. (eds) Advances in Visual Informatics. IVIC 2017. Lecture Notes in Computer Science, vol 10645. Springer, Cham, 2017, 175-184
[19] Rahm, Erhard, and Hong Hai Do. "Data cleaning: Problems and current approaches." IEEE Data Eng. Bull. 23.4 (2000): 3-13.

IOP Conf. Series: Journal of Physics: Conf. Series **1018** (2018) 012012 doi:10.1088/1742-6596/1018/1/012012

Human-Technology Centric In Cyber Security Maintenance For Digital Transformation Era

Firkhan Ali Bin Hamid Ali[1], Dr Mohd Zalisham Jali[2]

[1]Faculty of Computer Science & Information Technology
Universiti Tun Hussein Onn Malaysia
Johor, Malaysia
firkhan1977@yahoo.com

[2]Faculty of Science & Technology
Universiti Sains Islam Malaysia
Negeri Sembilan, Malaysia
zali@usim.edu.my

Abstract—The development of the digital transformation in the organizations has become more expanding in these present and future years. This is because of the active demand to use the ICT services among all the organizations whether in the government agencies or private sectors. While digital transformation has led manufacturers to incorporate sensors and software analytics into their offerings, the same innovation has also brought pressure to offer clients more accommodating appliance deployment options. So, their needs a well plan to implement the cyber infrastructures and equipment. The cyber security play important role to ensure that the ICT components or infrastructures execute well along the organization's business successful. This paper will present a study of security management models to guideline the security maintenance on existing cyber infrastructures. In order to perform security model for the currently existing cyber infrastructures, combination of the some security workforces and security process of extracting the security maintenance in cyber infrastructures. In the assessment, the focused on the cyber security maintenance within security models in cyber infrastructures and presented a way for the theoretical and practical analysis based on the selected security management models. Then, the proposed model does evaluation for the analysis which can be used to obtain insights into the configuration and to specify desired and undesired configurations. The implemented cyber security maintenance within security management model in a prototype and evaluated it for practical and theoretical scenarios. Furthermore, a framework model is presented which allows the evaluation of configuration changes in the agile and dynamic cyber infrastructure environments with regard to properties like vulnerabilities or expected availability. In case of a security perspective, this evaluation can be used to monitor the security levels of the configuration over its lifetime and to indicate degradations.

Keywords—cyber security; security maintenance; digital transformation; security management model.

IOP Conf. Series: Journal of Physics: Conf. Series **1018** (2018) 012012 doi:10.1088/1742-6596/1018/1/012012

Introduction

The development of Information Communication Technology (ICT) infrastructure had growth which is very fast in producing a wide range of computer products cause some medium sized organizations are confused and ambiguous as to what should be done to the ICT infrastructure. This resulted in tragedy 'white elephant' where infrastructure is purchased by the organization were not fully utilized or not used at all especially for ICT security infrastructure. This ambiguity is likely due to the lack of control or it does not give the impression and clear benefits to business activities and organizational management.

Thus, the ICT infrastructure needs of a medium-sized organizations will be discussed to overcome this problem is not clear. Plans are made to the infrastructure of ICT is not simply to facilitate commerce and organization but it must be in line with the mission and objectives of the organization.

Such planning should be in terms of ICT needs of an organization. Next, it seeks to be implemented and operated well for the success of an organization, whether it is for profit or social based service.

ICT security maintenance is very important aspect in ICT infrastructure to identify any weaknesses which involved security breach in the some organizations at early time. At the same time, some key concerns have also emerged about security maintenance in IT services and infrastructure, which currently are viewed as significant barriers to its fast and wide-spread adoption. According to an IDC survey of CIOs consecutively in 2008 and 2009, security, integration and reliable performance ranked among the top concerns expressed [1].

An ENISA (European Network and Security Administration) survey of Small and Medium Business (SMBs) also confirms that major concerns for SMBs migrating to the ICT service or infrastructure include the confidentiality of their information and liability for incidents involving the infrastructure [2]. This is understandable, because each of these factors have a major influence on the enterprises bottom-line.

Similarly, availability of the ICT services platform with good performance and depends heavily on the quality of network characteristics, especially the round trip delay or latency [1]. Security is a key concern, because confidentiality, integrity, authenticity and auditability of business data, tools and transactions are critical requirements for businesses to stay functional, legal and competitive. This need is especially critical for all users especially for overall security maintenance in IT service and infrastructure.

ISO network management model

Management of operational security maintenance in ICT infrastructure must adopt a management model such as ISO network management model. ISO network management model is a five functional area of network management that developed by The International Organization for Standardization (ISO). It provides a design guideline for future implementation of network management tools and technologies. Five functional area of network management consist of Performance Management, Configuration Management, Accounting Management, Fault management and Security Management.

Performance management functions in ISO network management model is to monitor, assessment and review of the available bandwidth and network usage of resources in a network to make more efficient to run. Performance management is a very important part of the ISO network management model, especially for ICT infrastructure that wants to streamline network performance in the organizations.

The goal of Configuration Management functions in ISO network management model is to monitor network and system configuration information so that the effects on network operation of various versions of hardware and software elements can be tracked and managed. It will centrally monitor configuration aspects of network devices such as configuration file management, inventory management and software management.

Accounting management functions in ISO network management model is the process to use in measuring network utilization parameters so that individual or group users on the network can be regulated appropriately for the purposes of accounting or chargeback. Same as the performance management, the first step toward appropriate accounting management is to measure the utilization of all important network resources. This aspect of network management usually focusing on Internet service providers to bill customers for the resources they use.

Fault management in ISO network management model is similar on what most people think that the administration thinks to manage the network. The goal of this network management functional is to identify, detect and alert system administrators of problems that may affect the system operations. Then, it needs to fix network problems primely in automatically to keep the network running effectively. Any faults can cause downtime or unacceptable network degradation, fault management is perhaps the most widely implemented of the ISO network management.

The goal of security management ISO network management model is to control access to network resources according to local guidelines so that the network cannot be sabotaged whether in intentionally or unintentionally. Implementation of security management subsystem it seems like, can monitor users logging on to a network resource, refusing access to those who enter inappropriate access codes. Security management deals with controlling access to resources. Then, notify the competent authorities if some resources are available. Similarly like a network operator or e-mail outsourcing, if a resource fails, management systems can be used to access the network to send messages when certain files or routers, servers.

It security management

The definition of Information Security based on ISO/IEC 17799:2005 is "preservation of confidentiality, integrity and availability of information, in addition, other properties such as authenticity, accountability, non-repudiation and reliability can also be involved".

Information has many definitions as follows:

i) Information is about someone or something consists of facts about them.

ii) Important or useful facts can be obtained as output from a computer by means of processing input data with a program.

iii) Information is an asset which is like other important business assets which is has value to an organization and consequently needs to be suitably protected [3].

iv) Information can exist in many forms. It can be printed or written on paper, stored electronically, transmitted by post or using electronic means, shown on films or spoken in conversation.

The core elements of information security management are to ensure the information assets, namely the following aspects.

i) Confidentiality
ii) Integrity
iii) Availability

Confidentiality is the property that information is not made available or disclosed to unauthorized individuals, entities, or processes [4]. Some examples of breach of confidentiality are "Unauthorized personnel can read the classified documents", "Remote access to the system without approval", and "Shared folders without consent of the owners".

Integrity separates into data integrity and system integrity [5]. Data integrity means the property that data has not been altered or destroyed in an unauthorized manner [6]. System integrity means the property that a system performs its intended function in an unimpaired manner, free from deliberate or accidental unauthorized manipulation of the system [7].

The definition of Availability is the property of being accessible and usable upon demand is obtainable from an authorized entity.

Security of Information Technology Infastructure Maintenance

Upon the successful implementation and testing of a new and improved security profile, an organization might feel more confident of the level of protection it is providing for its information assets [8]. By the time the organization has completed implementing the changes mandated by an upgraded security program, a good deal of time has passed.

In that time, everything that is dynamic in the organization's environment has changed [8]. Some of the factors that are likely to shift in the information security environment are:

- New assets are acquired.
- New vulnerabilities associated with the new or existing assets emerge.
- Business priorities shift.
- New partnerships are formed.
- Old partnerships dissolve.
- Organizational divestiture and acquisition occur.
- Employees who are trained, educated, and made aware of the new policies, procedures, and technologies leave.
- New personnel are hired possibly creating new vulnerabilities.

If the program is not adjusting adequately to change, it may be necessary to begin the cycle again. That decision depends on how much change has occurred and how well the organization and its program for information security maintenance can accommodate change [8]. If an organization deals successfully with change and has created procedures and systems that can flex with the environment, the security program can probably continue to adapt successfully.

The CISO determines whether the information security group can adapt adequately and maintain the information security profile of the organization or whether the macroscopic process of the SecSDLC (Security System Development Life Cycle) must start a new to redevelop a fundamentally new information security profile.

It is less expensive and more effective when an information security program is designed and implemented to deal with change [8]. It is more expensive to reengineer the information security profile again and again.

Management model must be adopted to manage and operate ongoing security program [9]. Models are frameworks that structure tasks of managing particular set of activities or business functions [10]. With that, by assist the information security community to manage and operate the ongoing security program, a management model must be adopted [11]. In general, management models are frameworks that structure the tasks of managing a particular set of activities or business functions.

A maintenance model is intended to complement the chosen management model and focus organizational effort on maintenance. This figure 1 diagrams a full maintenance program and forms a framework for the discussion of maintenance that follows.

- External monitoring
- Internal monitoring
- Planning and risk assessment
- Vulnerability assessment and remediation
- Readiness and review

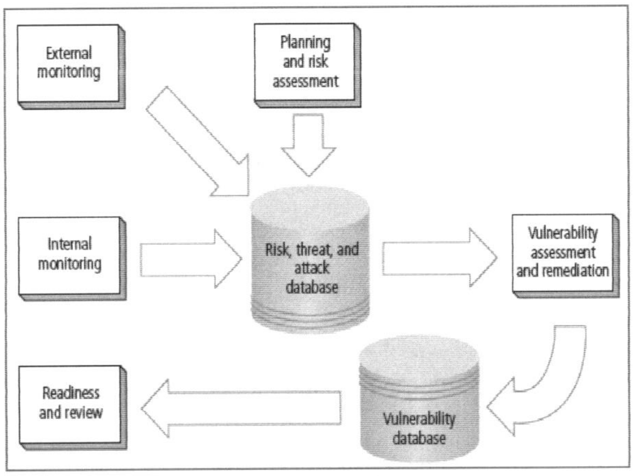

Fig.1. The maintenance model

Types of Security Threats And Risks

Generally, there are three types of risks that threaten the security of data in the form of digital environment or IT infrastructure which are vulnerability, threat and attack. In the beginning, all the vulnerability and threats are occurring passively in information technology infrastructure. However, it is necessary to solve and overcome this passively weaknesses and threats before it becomes as an active mode. The attack is represents as a risk that is occur in actively [12].

Further following discussion is about each type of risk that is capable of threatening the security of the information technology infrastructure. All the risks can go beyond the data security which is very valuable to an organization or company such as in Figure 2.

1. Vulnerability is anything that is weaknesses occur in used of information technology infrastructure and services. This vulnerability can be existing in many situations such as in design, configuration, implementation or management of information technology infrastructure that makes it vulnerable to the threats [13]. The downside is what makes information technology infrastructure vulnerable to the information loss and downtime issues or down time.

2. Threat in the cyber world is anything that can be interferes with the operation, function, integrity, availability of all types of IT infrastructure and services. Threats can be occur in any type of form. Threats can be happen as an evil action by those who are not responsible or by accident due to natural events or human errors.

3. Attack is the particular technique used to exploit the existing weaknesses that occur in information technology infrastructure [14]. For an example is a type of threat as a denial of service or DoS. Actually, this weakness exists in the design of operating systems and a type of attack that can be done due to the weaknesses is an attack called Ping of Death attacks. There are two main categories of attack which is passive attack or active attack.

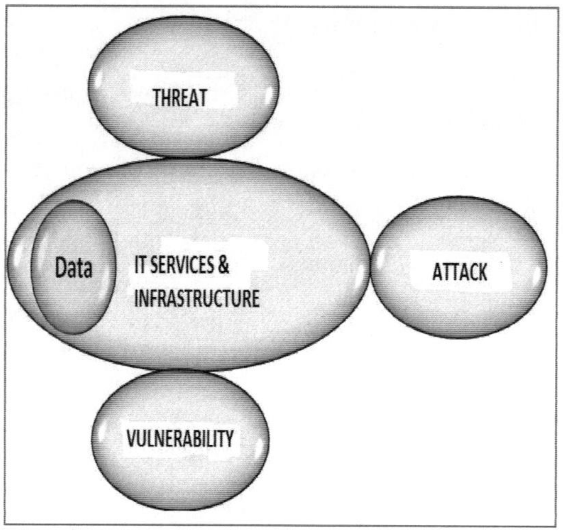

Fig. 2. IT Security Risks

Information security management model review

ICT infrastructure security is a state which protects a system or object from risk. It consists of systems, operations and internal controls in order to ensure the confidentiality, availability and integrity of data, knowledge and ICT infrastructure of an organization. User's roles in ICT systems have evolved from IT specialists for accessing information facilities, non-IT personnel for daily operation, unspecified individuals which is interested parties from outside organization. Establishment of organization information security policy (ISP) is the first step to protect organizations from security threats. Figure 3 summarized procedures for building a set of ISP through the comparison of six standards in information security management standard.

International standard	Guidelines
BS 7799 (Code of practice for ISM) ISO/IEC 17799	Describe the minimum contents in an ISP Explain what should do with an ISP An ISP should be approved by management, published and communicated throughout the organization An ISP should be evaluated and updated periodically
BSI IT baseline protection manual	Description of drawing up an ISP Coverage topics Contents Review
COBIT (The information system audit and control association and foundation, ISACAF)	Describes the process and control needs for implementing an ISP Brief section on the security and internal control framework policy Writing and maintaining a document
GASSP (Generally accepted system security principles)	Minimal requirements for an ISP and principles behind it The different processes needed for defining, maintaining and implementing the policy The hierarchy concept of an ISP
GMITS (ISO/IEC PDTR 13335-1)	Provide a comprehensive guidance on information security with planning, management and implementation
ISFs Standard of good practice (The globally representation information security forum, ISF)	Performance evaluation List the contents of ISP The characteristics of the policy Explain the acceptable user behavior

Fig.3. International standard for Information Security Management Model

As a pioneer, UK Department of Trade & Industry (DTI) firstly developed Code of Practice (CoP) PD0003 on information security in September 1993, with the assistance of a group of leading UK organizations. This Code of Practice was later retitled and published as BS 7799 Part 1 "Code of Practice for Information Security Management" in February 1995 by British Standards Institution (BSI). BS 7799 provides a common basis for developing organizational security standards and effective information security management practices. It enhances confidence in inter-organizational dealings.

Then, a new standard BS 7799 Part 2 "Information Security Management System – Specification with guidance for use" was released in 1998. The structure of this standard was the same as Part 1, in addition to defining a Code of Practice based on a set of key controls. As BS 7799 was a theoretical control standard and not a technical standard of practice, it might not solve ISMS problems effectively. Therefore, ISO further developed ISO/TR 13335 (Information Technology – Guideline for the Management of IT Security) and ISO/IEC 18044 (Information Security Incident Management) standards (ISO/IEC FDIS 17799:2005; ISO/IEC 27001:2005 [15]. Both standards are helping ICT industry to implement information security management.

A scheme for accreditation of BS 7799 entitled "c:cure" was launched at InfoSecurity 1998 by UK Accreditation Certification Service (UKAS) and the British Standards Institution (BSI). The accreditation procedure for ISO 9001 was adopted such as independent accredited certification body required for the purpose. This scheme initiated the further adaptation from national standard (BS) to international standard (ISO). Following the revisions of BS 7799 part 1 in 1999, the standard was transferred to ISO/IEC 17799:2000 (Part 1) – Code of Practice for Information Security Management. Finally, ISO/IEC 27001 Part 1 and Part 2 were issued in 2005, making them official and recognized standards both locally and internationally [16]. ISO/IEC 27001:2005 is directly related to the original BS 7799.

ISO/IEC 27002:2005 is a generic and advisory document, not a formal specification standard. It provides a well-structured and comprehensive set of controls to address information security risks, covering confidentiality, integrity and availability aspects [17]. Organizations that adopt ISO/IEC 27002 must assess their own information security risks and apply suitable controls, by following the standard for guidance.

ISO 27001

ISO 27001 is an International Standard which specifies the requirements for establishing, implementing, operating, monitoring, reviewing, maintaining and improving a documented Information Security Management System (ISMS) of an organization's overall business risks. ISMS is designed to ensure the selection of security controls is adequate and proportionate so as to protect information assets of the organization and give confidence to their customers.

ISO/IEC 27001:2005 (Information Security Management System - Requirements)

The international standard ISO/IEC 27001:2005 has its roots in the technical content derived from BSI standard BS7799 Part 2:2002. This standard is generally applicable to all types of organizations, including business, enterprises, government agencies, institution, healthcare and so on.

The standard introduces a cyclic model known as the "Plan-Do-Check-Act" (PDCA) model that aims to establish, implement, monitor and improve the effectiveness of an organization's ISMS. The PDCA cycle has four phases:

1. Plan – Establishment of the ISMS. The first step is to define risk assessment, in which risks shall be identified, analyzed and evaluated. Identification and evaluation of the risk treatment options are then followed. After the control objectives and controls are selected, management needs to approve residual risks and authorize implementation of ISMS.
2. Do – Implementation and operating the ISMS. Management actions, resources, priorities, roles and responsibilities shall be defined in this step. It needs to determine the risk treatment plan to the respective risks, and to implement controls accordingly.

IOP Conf. Series: Journal of Physics: Conf. Series **1018** (2018) 012012 doi:10.1088/1742-6596/1018/1/012012

3. Check – Monitoring and reviewing the ISMS. Monitoring and reviewing procedures should be developed and executed. The effectiveness of ISMS and controls, as well as the risk assessment methodology and residual risks, should be included and reviewed.
4. Act – Maintaining and improving the ISMS. Implementing both preventive and corrective actions in this step could further improve the ISMS. It also enforces document and record control, and reviews information security incidents for lessons learning, so as to improve the ISMS.

Often, ISO/IEC 27001:2005 is implemented together with ISO/IEC 27002:2005. ISO/IEC 27001 and ISO/IEC 27002 assist in defining the requirements and outlining the most suitable information security controls for the ISMS respectively. However, no guidelines of risk assessment mechanism and ISMS implementation are included; those standards are only stated what is needed to do but it does not mention how to do.

ISO/IEC 27002:2005 (Code of Practice for Information Security Management)

ISO/IEC 27002:2005 (replacing ISO/IEC 17799:2005 in April 2007) is an international standard that originated from the BS7799-1 standard originally laid down by the British Standards Institute (BSI) [18]. ISO/IEC 27002:2005 refers to a code of practice for information security management, and is intended to be a common basis and practical guideline for developing organizational security standards to facilitate the effectiveness of management practices.

By adopting the standard, an organization can address information security risks comprehensively [19]. This standard consists of several guidelines and the best practices in 11 security domains shown as follows:

1. Security policy.
2. Organization of information security.
3. Asset management.
4. Human resources security.
5. Physical and environmental security.
6. Communications and operations management.
7. Access control.
8. Information systems acquisition, development and maintenance.
9. Information security incident management.
10. Business continuity management.
11. Compliance.

Among these 10 security domains, a total of 39 control objectives and 133 best practice information security control measures are recommended for organizations to satisfy the control objectives and protect information assets against threats to confidentiality, integrity and availability.

ISO/IEC 13335 (IT Security Management)

ISO/IEC 13335 was initially a Technical Report (TR) before becoming a full ISO/IEC standard. It consists of a series of guidelines for technical security control measures:

1. ISO/IEC 13335-1:2004 documents the concepts and models for information and communications technology security management.

2. ISO/IEC TR 13335-3:1998 provides the techniques for the management of IT security. It was superseded by ISO/IEC 27005:2008.

3. ISO/IEC TR 13335-4:2000 covers the selection of safeguards (for example technical security controls). It was superseded by ISO/IEC 27005:2008.

4. ISO/IEC TR 13335-5:2001 suggests management guidance on network security. It has been under review, and may merge with ISO/IEC 18028-1 and ISO/IEC 27033.

Information security is a multi-dimensional discipline that can be viewed from different perspectives. The conceptual models are presenting security element relationship that shows how assets are potentially subjected to a number of threats. The threats and environment change over time and it may have impacts on the probability of risk occurrence [20].

The model has been developed with the assumption of an environment containing constraints and threats that change constantly and are only partially known, including the assets of an organization, the vulnerabilities associated with those assets, safeguards selected to protect assets and residual risks acceptable to the organization [21].

Methodology & proposed framework

The methodology of the proposed research will be carried out based on the fundamental of the experimental information technology method. This method examines the research work to demonstrate two important concepts: proof-of-concept and proof-of-performance.

To demonstrate the proof-of-concept, some important steps were performed. First, the research area within security maintenance is critically reviewed to provide the overview that leads to the justification of a valid research problem. Then, a novel model of the security maintenance framework is designed and analytically analysed. This includes the creation of the mechanism for managing security model, processes and metrics in relation to use of security maintenance.

Proof-of-performance is demonstrated by integrating the proposed security model, processes and metrics within a novel conceptual framework of the security maintenance in IT infrastructure. Then, it will be assessed using proposed framework. In those proposed framework, various parameters and workloads were used to examine and demonstrate the viability of the proposed solutions compared to other similar baseline solutions. Also, analytical analysis of some proposed security metrics is performed to evaluate the correctness.

Specifically, the main stages involved in the research are divided into three:

- Data Acquisition Stage
- Investigation and Modelling Stage
- Analysis and Evaluation Stage

The proposed conceptual framework for security maintenance in mid-size IT infrastructure represents as an integration of IT security management model and concepts covering several facets of information security aspects and processes. The framework draws from multiple areas including software vulnerability, risk assessment, attack motivation, threat detection, deterrence and security objective [22]. It is based on an earlier model for information security management model. The framework has been enhanced with the inclusion of combination of constructs and refined through the recalibration of IT security management model to ensure that potentially anomalous situations are prevented. The proposed framework is depicted as in Figure 4.

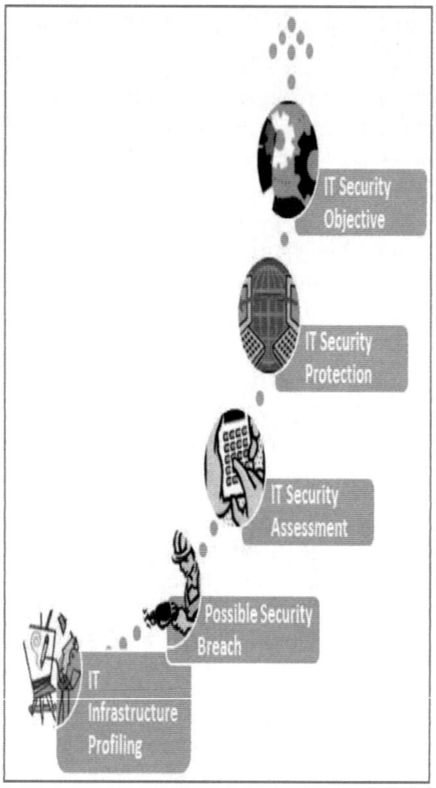

Fig. 4. Proposed IT Security Maintenance Framework

Conclusion

After review of information security management, this provides foundation knowledge and supporting evidences to develop a security maintenance model for ICT infrastructure. The development on Information Security Management System (ISMS) had a long development history in any organization and low adoption of ISO 27001 was observed [23]. The high costs in money and time of ISMS implementation are definite barriers for smaller size companies to adopt the standard.

Therefore, a security maintenance model for ICT infrastructure based on putting the core and compatible requirements of ISO 27001 ISMS for reducing the redundancy of the existing usage of ICT security safeguards and aimed to reduce domain's barriers [24]. The development of a security maintenance model for ICT infrastructure is proposed as a common framework and overcome the limitations of integrated system theory, as well as adopting the key concepts from IT security model.

Moreover, the PDCA approach was observed in ISO 27001 ISMS model. Combination of all the advantages in each model as example like understanding customer requirements, value-added processes, processes performance and effectiveness, continually improvement and others to develop the security maintenance model for ICT infrastructure.

A review on different approaches of IT security management model was performed such as ISO/IEC 27005:2008 and PDCA framework. This is combined with PDCA and IT security management model to fulfill ISO 27001:2005 standard. Lastly, the different implementation models for ISMS and IT security management model had reviewed.

Actually, a novel conceptual framework of the security maintenance had proposed to make any IT infrastructures and services that follow those guidelines will accessible properly and secure by any authorized peoples. However, there are no security aspects had been discuss in details for IT services and infrastructure's maintenance. Then, the documentation had been provided as a manual access which is needs to follow the guidelines.

Now days, security should be concern in any IT services and infrastructures including in any proposed maintenance model and guidelines. Then, its need overall coverage to make the guidelines become more effective and easy to use. Security maintenance is more important in cyber space for any organizations especially for IT services and infrastructure usage in safe and secure manner.

Acknowledgment

Thanks for the Universiti Tun Hussein Onn Malaysia on the supporting to this paper and make it available in the proceeding.

References

[1] J. Fonseca, M. Vieira and H. Madeira, "Evaluation of Web Security Mechanisms using Vulnerability and Attack Injection," IEEE Transactions on Dependable and Secure Computing, vol. PP, no. 99, pp. 1-1, 2013.

[2] Wooyun. wooyun, <http://www.wooyun.org/> [accessed 01.05.15].

[3] C. Melara, J.M. Sarriegui, J.J. Gonzalez, A. Sawicka, D.L. Cooke, A system dynamics model of an insider attack on an information system, in: J.J. Gonzalez (Ed.), From Modeling to Managing Security: a System Dynamics Approach, Norwegian Academic Press, Kristiansand, Norway, 2003, pp. 9–36.

[4] C.E. Landwehr, Formal models for computer security, ACM Comput. Surv. 13, 1981, pp. 247–278.

[5] 360. Butian, <http://loudong.360.cn/> [accessed 01.05.15]

[6] D.L. Nazareth, J. Choi, Information security management: a system dynamics approach, Eighteenth Americas Conference on Information Systems (AMCIS- 2012), Seattle, WA, 2012, Paper 3.

[7] D. Trcek, Using system dynamics for managing risks in information systems, WSEAS Trans. Inf. Sci. Appl. 2, 2008, pp. 175–180.

[8] M.E. Whitman, H.J. Mattord, Principles of Information Security, fourth ed., Course Technology, Boston, MA, 2014.

[9] M.A. Alnatheer, A conceptual model to understand information security culture, Int. J. Soc. Sci. Hum. 4, 2014, pp. 104–107.

[10] J. May, Analyzing the socio-organizational constructs for IS security within organizations, in: S. Furnell, P. Dowland (Eds.), in: Proceedings of the 11th IFIP TC11. 1 Working Conference on Information Security Management, Richmond, VA, 2008, pp. 103–118.

[11] K.H. Guo, Y. Yuan, The effects of multilevel sanctions on information security violations: a mediating model, Inf. Manage. 49, 2012, pp. 320–326.

[12] A. Simmonds, P. Sandilands, L.v. Ekert, An ontology for network security attacks, in: S. Manandhar, J. Austin, U. Desai, Y. Oyanagi, A. Talukder (Eds.), Applied Computing, Springer, Berlin, 2004, pp. 317–323.

[13] D. Trcek, Security models: refocusing on the human factor, Computer 39, 2006, pp. 103–104.

[14] M.d. Vivo, G.O.d. Vivo, G. Isern, Internet security attacks at the basic levels, ACM SIGOPS Oper. Syst. Rev. 32, 1998, pp. 4–15.

[15] ISO/IEC 17799 (2005), Information Technology – Security Techniques – Code of Practice for Information Services, International Organization for Standardization, Geneva.

[16] ISO/IEC 17799 (2005), Information Technology – Security Techniques – Code of Practice for Information Services, International Organization for Standardization, Geneva.

[17] ISO/IEC 27001 (2005), Information Technology – Security Techniques – Information Security Management System – Requirements, International Organization for Standardization, Geneva.

[18] ISO/IEC 27002 (2005), Information Technology – Security Techniques – Code of Practice for Information Services, International Organization for Standardization, Geneva.

[19] Tripathi A, Singh UK. Taxonomic analysis of classification schemes in vulnerability databases. In: Computer sciences and convergence information technology (ICCIT), 2011 6th international conference on. IEEE; 2011. p. 686–91.

[20] US-CERT. Common vulnerabilities and exposures (cve), <https://cve.mitre.org/> [accessed 01.05.15].

[21] A. Hovav, J. D'Arcy, Applying an extended model of deterrence across cultures: an investigation of information systems misuse in the U.S. and South Korea, Inf. Manage. 49, 2012, pp. 99–110.

[22] S.-C. Yang, Y.-L. Wang, Insider threat analysis of case based system dynamics, Adv. Comput. 2, 2011, pp. 1–17.

[23] A.C. Kim, S.M. Lee, D.H. Lee, Compliance risk assessment measures of financial information security using system dynamics, Int. J. Secur. Appl. 6, 2012, pp. 191–200.

[24] A. Vance, M. Siponen, S. Pahnila, Motivating IS security compliance: insights from habit and protection motivation theory, Inf. Manage. 49, 2012, pp. 190–198.

IOP Conf. Series: Journal of Physics: Conf. Series **1018** (2018) 012013 doi:10.1088/1742-6596/1018/1/012013

The Prospect of Internet of Things and Big Data Analytics in Transportation System

Waleed Noori Hussein[1,a], L.M. Kamarudin[2,3,b], Haider N.Hussain[4,c],A. Zakaria[2,d] ,R Badlishah Ahmed[3,e] , N.A.H.Zahri[3,f]

[1] School of Human Development and Technocommunication
[2]Centre of Excellence for Advanced Sensor Technology (CEASTech)
[3]School of Computer and Communication Engineering
Universiti Malaysia Perlis (UNIMAP),
02600 Arau, Perlis, Malaysia
[4]Faculty of Education for Pure Science
University of Basra
[a]waleedn9@yahoo.com, [b]latifahmunirah@unimap.edu.my , [c]alhashmiy@gmail.com,
[d]ammarzakaria@unimap.edu.my, [e]badli@unisza.edu.my,
[f]adilahhanin@unimap.edu.my

Abstract. Internet of Things (IoT); the new dawn technology that describes how data, people and interconnected physical objects act based on communicated information, and big data analytics have been adopted by diverse domains for varying purposes. Manufacturing, agriculture, banks, oil and gas, healthcare, retail, hospitality, and food services are few of the sectors that have adopted and massively utilized IoT and big data analytics. The transportation industry is also an early adopter, with significant attendant effects on its processes of tracking shipment, freight monitoring, and transparent warehousing. This is recorded in countries like England, Singapore, Portugal, and Germany, while Malaysia is currently assessing the potentials and researching a purpose-driven adoption and implementation. This paper, based on review of related literature, presents a summary of the inherent prospects in adopting IoT and big data analytics in the Malaysia transportation system. Efficient and safe port environment, predictive maintenance and remote management, boundary-less software platform and connected ecosystem, among others, are the inherent benefits in the IoT and big data analytics for the Malaysia transportation system.

Keywords: Internet of Things; Big Data Analytics; Malaysia transportation systems

1. Introduction

The Malaysia transportation system has potential benefits from harnessing the strengths and functionalities of Internet of Things (IoT), and applying the analytics of its big data. IoT is defined as the process and technological framework that explains the interaction, interconnection and interdependence among data, people, and associated electronic objects over the internet [1-3]. IoT is also aided by the sensory and communication abilities of the connected physical electronic objects. Through these objects, the environment is monitored and reported, and communicating electronic objects are programmed to act according to the information received [1, 4, 5]. Different varieties, large volume, and high velocity data (big data) are generated due to the intercommunication among the physical electronic objects. Therefore, exploring the analytics of the generated big data, for prescription, description, prediction, and actionable insights become necessary.

IoT and big data analytics have become the dawning technological revolution, and due to its enormous benefits, they have been adopted in diverse domains for varying purposes. IoT and big data analytics have been utilized in manufacturing, agriculture, banks, oil and gas, healthcare, retail, hospitality, food services, among others [6-8]. Automated and robotic manufacturing [9, 10] automated teller machine's fraud detection [11, 12], and human profile security alert [13, 14] are applications of IoT and big data analytics recorded in manufacturing, banking, and security, respectively, to mention a few. The transportation industry is also one of the early adopters of IoT and big data analytics [6, 7]. Its applications in tracking shipment, freight monitoring, and transparent warehousing, signify the benefits of IoT and big data analytics in the transportation industry as recorded in the transportation sector of countries like England, Germany, Portugal, Singapore [7], among others. However, Malaysia can be said to still be at the teething stage.

This paper presents a summary of the inherent prospects in adopting IoT and big data analytics in the Malaysian transportation system. In doing this, the next section of this paper describes the general overview of the applications of IoT and big data analytics in transportation system. Countries that have adopted IoT and big data analytics in their transportation systems are presented, and the impact of its adoption are highlighted. In the third section, the present situation of Malaysia transportation system is described, and the prospects of adopting IoT and big data analytics are enumerated. The fourth section is the last and concluding part of this paper.

2. Applications of Internet of Things and Big Data Analytics in Transportation System

The transportation sector is in the forefront of IoT technologies implementation and it is rapidly adopting its innovations to remain competitive, increase productivity and bottom-line, and to enhance customer loyalty and satisfaction. This is much necessary because of the evidently low-margin fragmentation that is typical of the transportation industry. IoT has been responsible for the timely need of the supply chain transparency, and its support to data-driven decision making in a connected and distributed network [6].

The track-and-trace global positioning system (GPS) technology is an IoT technology that was firstly used in tracking shipment before the emergence of radio-frequency identification (RFID) technology. Now, virtually the entire logistics chain warehousing, freight transportation, and last mile delivery have adopted IoT technology, with an estimated €16.8 trillion investment stake [15]. As shown from literature, Portugal [16], Germany [6], England [17], and Singapore [18] are some of the leading countries that have massively adopted IoT and the associated big data analytics for revolutionizing their transport systems.

In Lisbon, Portugal, according to Baptista, Azevedo [16], there was a problem of non-compliance with speed limits by majority of the drivers, resulting in recurring cases of fatal car accidents. This portends negative image for the national transportation agency and the expected safety policy. It is against this backdrop that IoT and big data analytics were employed. A study of the potential environmental and economic impacts of implementing information communication technological (ICT) measures in transportation was conducted. The implementation of IoT, big data analytics, and other associated ICT applications achieved a behavioral change in the drivers' eco-driving and public decorum attitudes. Subsequently, a drastic reduction in the cases of fatal car accidents in Lisbon was achieved. Also, there was considerable impact in eco-friendly environment in terms of reduced energy consumption and CO_2 emission.

In the same vein, the Port of Hamburg in Germany being one of the busiest ports in Europe was facing traffic rise which was compounded by its space limitation. There are other factors, such as road safety concern, traffic congestion, and pollution, which must be addressed in view of creating safe and low-cost port environment. The port is also a point for the tourism industry because of the large number of cruise passengers that use it [19]. To these ends, IoT technologies were employed and strategically implemented. Smart port infrastructure, intelligent trade flow, and intelligent traffic flow are the main sub-components of the IoT technological framework adopted. As a result, Port of Hamburg became a major economic driver for the country and the region by creating more than 260 thousand jobs and generating over €750 million as yearly tax revenue. With the capacity of the IoT technologies, the port now handles container and bulk cargo which are expected from 2010's 8 million to 25 million in 2025 [20].

There was also the need for operational waste reduction, timely execution of project and worker/passenger safety in the England's transportation sector. Tezel [17] explored both the quantitative and qualitative benefits of a practical visual system developed with IoT technologies. The implementation of the system contributed to increased self-management, better team coordination, and more efficient delivery of cargo. It also helped in easier control of the workforce and their activities in the transportation sector.

In Singapore, also, there was a dire need for consumers' choice to evaluate different bus option in view of making wise purchase decision. IoT and big data analytics infrastructure were therefore employed for better passengers' experience in bus transportation. The industry experts were also interviewed in order to understand the necessities in the domain and how IoT can be explored for better and more efficient transportation service delivery in Singapore [18]. The study produced an IoT-based technical architecture for an application that predicts arrival times for the buses, makes comparative analysis of passengers' options, and gives estimate of the number of passengers in the arriving buses. This ultimately guides the passengers in choosing their preferred bus and route.

The benefits in the implementation of IoT and big data analytics in the transportation system cannot be overemphasized. Portugal, Germany, England and Singapore have been able to overcome the problems of fatal car accidents, traffic rise, operational waste, and passengers' decision making using IoT and big data analytics applications. These prospects are also achievable in Malaysian transportation system with careful study of the sector and strategic implementation of the applicable IoT technologies and big data analytics. Table 1 presents the problems observed and solved in these countries with IoT and big data analytics.

Table 1: The Summary of IoT and Big Data Analytics Implementation in Transportation Sector

Countries	Problems (before the implementation of IoT and Big data analytics)	Results (after the implementation of IoT and Big data analytics)
Portugal	Recurring cases of fatal accidents due to non-compliance with speed limits	Behavioral change in the drivers' eco-driving, then drastic reduction in the cases of fatal car accidents
Germany	Road safety concern, traffic congestion, and pollution.	Eco-friendliness, and hitch-free traffic flow
England	Operational waste and workers' safety	Operational waste reduction and safe working environment
Singapore	Unpleasant passenger's booking experience	Better passenger's booking experience

3. Malaysia Transportation System and the Prospects of Internet of Things (IoT) and Big Data Analytics Implementation

The prospects of IoT implementation and big data analytics applications in Malaysian transportation system lie in their strengths in solving the myriads of problems in the sector. This will also address the externally-influenced issues in the industry, and finally support the modernization of the entire system.

Similar to what were experienced in Portugal, Germany, England, and Singapore, Malaysia can also move into a top-notch service delivery country in its airport, train, seaport, and city bus transportation.

According to *Transport Statistics Malaysia*, 2015, Malaysian international and domestic airport facilities operate 24 hours, with total passengers excluding transit passengers of 85, 948, 179 within the year 2006 and 2015. The airport facilities also handled cargoes of 959, 042 metric tonnes, and 857, 232 mails within the same year span. There is also remarkable freight traffic of 1, 474, 35 as the number of registered vehicles on Malaysia roads within the year 2006 and 2015. The Malaysia transport capacity for the year 2006 – 2015 is depicted in Figure 1. This figure indicate that Malaysia transportation sector is thriving with massive international and local demands to meet. This, undoubtedly, comes with its problems and challenges. The leading among these problems, as indicated by *Transport Statistics Malaysia*, 2015, is road accidents, which are 489,606 cases in 2015 [21]. Though, the causes are not documented. Also of important attention are the operational bottlenecks, environmental hazard, such as noise pollution, traffic congestions, passengers' choice conflict, among others associated issues of growing modern transportation sector.

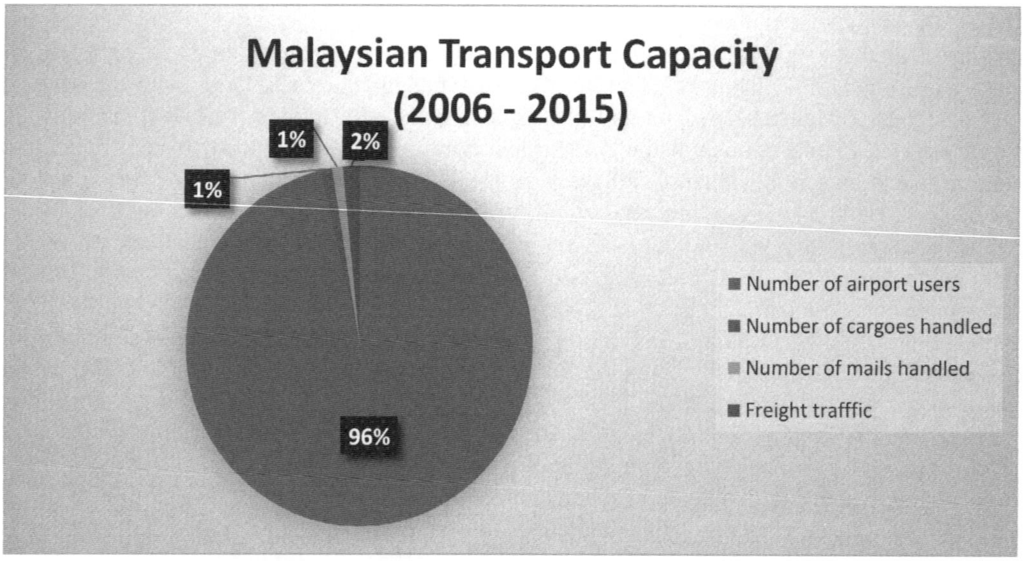

Figure 1. Malaysian Transport Capacity (2006 – 2015)

The lessons in the application of IoT technologies and big data analytics to reduce high rate of fatal car accidents in Lisbon, Portugal, can equally be applied in Malaysia. Most importantly is a generic overview of the applications of IoT and big data analytics in Portugal, Germany, England and Singapore. From this, Malaysia tends to achieve (a) reduction in cases of road accidents, (b) eco-friendly and user-friendly transportation environment, (c) revenue generation, (d) job creation, and (e) enhanced productivity and efficient service delivery, by applying IoT and big data analytics in its transportation system.

4. Conclusion

There are massive benefits in the adoption and implementation of IoT technologies and big data analytics. This is evident in its diverse adoption across domains, such as hospitality, oil and gas, security, healthcare, among others, and transportation, as focused in this paper. Countries like Portugal, Germany, England and Singapore have taken the front row in the application of IoT and big data analytics to solve varying problems in their transportation sectors. These problems range from incessant fatal car accidents, ports' traffic rise and congestion, operational waste, and bus passengers' choice.

This paper presents the prospects of applying big data analytics and implementing IoT technologies in the Malaysia transportation sector. Asides the generic benefits of port safety, supply chain transparency, and its support to data-driven decision making in a connected and distributed network, there are Malaysian peculiarities which will be ultimately addressed. These are closely related with what are obtained in the transportation sectors of the countries reviewed, while the Malaysia sector also need tailored study.

Notably, IoT and big data analytics are offshoots of the broader Industrial Internet which must be paid adequate attention. These technologies are disrupting business model and modes of service delivery. Technology providers, technology adopters, public policy makers, and all stakeholders in the Malaysia transportation sector must be ready to work in synergy. There is need to raise market awareness on successful use cases, identify partners in the sector's technological ecosystem, revisit industries' regulations and policies, and implement new training model for the sector's workforce. These will cumulatively actualize the prospects of IoT and big data analytics for the transportation sector.

References

[1] Wang, X. and Z. Li, *Traffic and Transportation Smart with Cloud Computing on Big Data.* IJCSA, 2016. **13**(1): p. 1-16.

[2] Calado, J.M.F., L.A. Osório, and R. Prata. *An Adaptive IoT Management Infrastructure for EcoTransport Networks*. in *Working Conference on Virtual Enterprises*. 2015. Springer.

[3] Zhou, H., B. Liu, and D. Wang, *Design and research of urban intelligent transportation system based on the internet of things.* Internet of Things, 2012: p. 572-580.

[4] Diez, M., C. Ott, and S. Weber, *Business Models for the Internet of Things.* 2016.

[5] Leng, Y. and L. Zhao. *Novel design of intelligent internet-of-vehicles management system based on cloud-computing and internet-of-things*. in *Electronic and Mechanical Engineering and Information Technology (EMEIT), 2011 International Conference On*. 2011. IEEE.

[6] Thulesius, M., *Digital Transformation of Ports: A Status of the Port of Hamburg and the Port of Singapore.* 2016.

[7] Agenda, W.E.F.I., *Industrial Internet of Things: Unleashing the Potential of Connected Products and Services.* a White Paper publication in collaboration with Accenture, 2015.

[8] Ogbuokiri, B., C. Udanor, and M. Agu, *Implementing bigdata analytics for small and medium enterprise (SME) regional growth.* Department of Computer science, University of Nigeria, Nsukka, Enugu state, 2015.

[9] Kara, D., & Carlaw, S (2014). , *The Internet of Robotic Things ABI Research.Retrieved from https://www.abiresearch.com/market-research/product/1019712-the-internet-of-robotic-things/ on 29th August, 2017.*

[10] Berbon, P., & Watkin, N. J (2010). , *The Industrial Internet: Robotics, Automation, and the Future of Manufacturing. Future Watch Report.Retrieved from https://www.tekes.fi/globalassets/julkaisut/cm_the_industrial_internet_automation_robotics_and_the_future_of_manufacturing_v13.pdf on 29th August, 2017.* .

[11] Menaga, S., Yamili, A., Rekha, P., & Tamilarasi, R. , *Internet of Things Based ATM Secure Monitoring.* International Journal of Innovative Research in Computer and Communication Engineering, 2017. **5**(3): p. 4712- 4717.

[12] Rajmohan, C., Raghavi, M., & Malavika, E. 5 (3), *ATM Theft Detection and Prevention Using IoT.* International Journal of Science and Engineering Research (IJ0SER), 2017. **5**(3).

[13] *US Homeland Security (2016), Strategic Principles for Securing Internet of Things (IoT).Retrieved from https://www.dhs.gov/sites/default/files/publications/Strategic_Principles_for_Securing_the_Internet_of_Things-2016-1115-FINAL....pdf on 29th August, 2017.* .

[14] Blowers, M., *Evolution of cyber technologies and operations to 2035.* 2015: Springer.

IOP Conf. Series: Journal of Physics: Conf. Series **1018** (2018) 012013 doi:10.1088/1742-6596/1018/1/012013

[15] Clarke, R.Y., *Smart cities and the internet of everything: The foundation for delivering next-generation citizen services.* Alexandria, VA, Tech. Rep, 2013.

[16] Baptista, P.C., I.L. Azevedo, and T.L. Farias, *ICT solutions in transportation systems: estimating the benefits and environmental impacts in the Lisbon.* Procedia-Social and Behavioral Sciences, 2012. **54**: p. 716-725.

[17] Tezel, A., et al., *Benefits of visual management in construction: cases from the transportation sector in England.* Construction Innovation, 2017. **17**(2): p. 125-157.

[18] Menon, A. and R. Sinha, *Implementation of Internet of Things in Bus Transport System of Singapore.* 2013.

[19] Ferretti, M. and F. Schiavone, *Internet of Things and business processes redesign in seaports: The case of Hamburg.* Business Process Management Journal, 2016. **22**(2): p. 271-284.

[20] Authority, H.P., *(2012).The Port Development Plan to 2025,http://www.hamburg-port-authority.de/de/presse/broschueren-und-publikationen/Documents/port-development-plan2025.pdf.*

[21] Statistics, M.T., *A Publication of Ministry of Transport, Malaysia.* 2015.

Aquaponic Growbed Water Level Control Using Fog Architecture

#1Muhamad Asmi Romli, #2Shuhaizar Daud, #3Rafikha Aliana A. Raof, #4Zahari Awang Ahmad and #5Norfadilla Mahrom

#School of Computer and Communication Engineering, Universiti Malaysia Perlis, Perlis, Malaysia
1asmi@unimap.edu.my
2shuhaizar@unimap.edu.my
3rafikha@unimap.edu.my
4zahari@unimap.edu.my
5norfadila@unimap.edu.my

Abstract— Integrated Multi-Trophic Aquaculture (IMTA) is an advance method of aquaculture which combines species with different nutritional needs to live together. The combination between aquatic live and crops is called aquaponics. Aquatic waste that normally removed by biofilters in normal aquaculture practice will be absorbed by crops in this practice. Aquaponics have few common components and growbed provide the best filtration function. In growbed a siphon act as mechanical structure to control water fill and flush process. Water to the growbed comes from fish tank with multiple flow speeds based on the pump specification and height. Too low speed and too fast flow rate can result in siphon malfunctionality. Pumps with variable speed do exist but it is costly. Majority of the aquaponic practitioner use single speed pump and try to match the pump speed with siphon operational requirement. In order to remove the matching requirement some control need to be introduced. Preliminarily this research will show the concept of fill-and-flush for multiple pumping speeds. The final aim of this paper is to show how water level management can be done to remove the speed dependency. The siphon tried to be controlled remotely since wireless data transmission quite practical in vast operational area. Fog architecture will be used in order to transmit sensor data and control command. This paper able to show the water able to be retented in the growbed within suggested duration by stopping the flow in once predefined level.

Keywords— Internet of Thing, IoT, Cloud, Automation, Remote, Monitoring, Control, Aquaponics, IMTA, MQTT

1. Introduction

Internet of Thing (IoT) is the inter-networking of physical devices, vehicles, buildings, and other items embedded with electronics, software, sensors, actuators, and network connectivity which enable these objects to collect and exchange data [1]. This technology evolves with the emerging internet capabilities. The introduction of this technology connects anything to the internet, enabling data and insights that never available before. According to Cisco Visual Networking Index [2], by 2020 46% of internet access will be represented by Machine-to-Machine (M2M) connection.

IoT works based on currently trending cloud computing. It is a type of internet-based computing that provides shared computer processing resources and data to computers and other devices on demand [3]. There are many companies provide cloud service with the names like Amazon, Microsoft Azure, IBM Watson, Google Cloud, Oracle, Salesforce, Bosch, Cisco, GE Predix and Thingspeak – just to name some of them. The introduction of this cloud consequently enables the generation of unprecedented numbers and variety of data. However, by the time the data available on the cloud for analysis, the opportunity to act on it might be gone. It seems like the cloud is suitable only for reference source.

Fog architecture has been introduced in order to avoid vast underused data to be uploaded to the cloud. This architecture seems to be more suitable to be implemented in data acquisition and control system like manufacturing, modern agriculture and aquaculture sector. The implementation also can remove the need for tedious wiring requirement especially in vast control area or remote location.

Modern aquaculture like IMTA or specifically aquaponics is currently getting more interest and adding automation will upscale the operation efficiency. In aquaponics, the implementation of growbed method seems to replace the need for expensive biofilter and drum filter. According to [4] this method provides high aeration and can support all types of plant. However the flood-and-drain mechanism in plant tanks requires careful calculation of water volume. Normal and affordable single speed pump specification affected by this requirement. This paper will show how to control flood-and-drain operation of a growbed by water level control method without the need to depend on water flow speed.

2. Background research

Aquaculture area nowadays in majority mostly is practiced in traditional way. Mechanical structure and human related operation widely used. The operational efficiency always in doubt when there is no automation involved. To eliminate this error-prone process, this paper will focus in automatic test which involve more on the electronic data acquisition and analysis. Three related elements of study had been carried out prior to the implementation of fog computing in aquaponics. Those three elements are fog computing, aquaponics and aquaponic growbed.

A. Fog Computing

Fog computing was introduced after the introduction of cloud computing. Edge computing or Fog computing is a new way of data sharing intended for IT and operational technology professionals. This is a new model which performs instant gathering and action on IoT data. Fog computing offers reduction of network traffic by providing a platform to filter and analyse the data generated by the devices which close to the edge and for local data views [5]. As the result, it will automatically reduce the traffic being sent to the cloud.

Fog computing suitable for IoT tasks and queries too where most of the smart devices need to capture events within hundred meters distance and no need to access global data from cloud. Low latency requirement also provided by fog computing which critical parameter can be reflected with high speed real time response. Furthermore, it can reduce the scalability issue since fog computing aims the incoming data to get closer to the data source itself and reduces the burden of that processing on the cloud and without concern of the increasing number of endpoints. Fog in actual is not a separate architecture; it merely extends the existing cloud architecture to the edge of the network—as close to the source of the data as possible—to enable real-time data processing and analytics [6]. Figure 1 shows how fog computing extends cloud capabilities to the edge of the network.

IOP Conf. Series: Journal of Physics: Conf. Series **1018** (2018) 012014 doi:10.1088/1742-6596/1018/1/012014

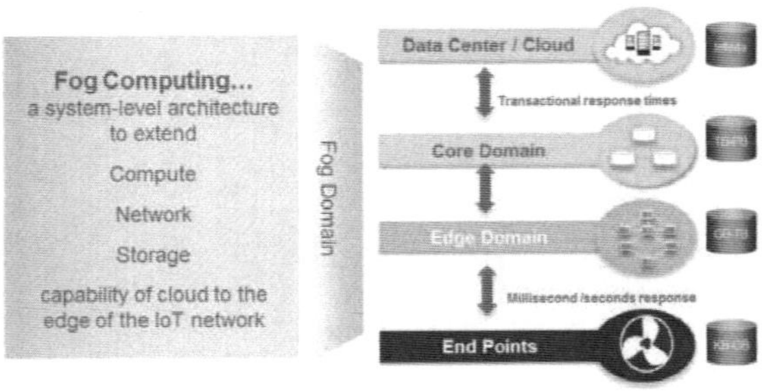

Figure 1: Cloud capabilities extension

B. Aquaponics

The word aquaponic comes from aquaculture and hydroponic. It is the integration of recirculating aquaculture system (RAS) and hydroponics in one closed production system [7]. The concept of by-product recycles happen here – this is why this system is categoried as Integrated Multi-Trophic Aquaculture (IMTA). The fish or aquatic animal produces waste and the crop system treat it as nutrient and absorb it. The clean water then return back to the tank. This is actually how aquaponic RAS works. In order to success in operating aquaponic, the RAS should be at the best performance. Inefficiency in RAS will result in catastropic lost especially to the aquatic life.

Nutrient Recovery Efficiency (NRE) is affected by many factors. By using this concept of IMTA, the variable can be limited and easily managed and hence can produce greater result. The source of nutrient purely comes from fish ponds or tanks, and this variable can be said to be constant. The one need to be altered in order to maximise NRE is the crop side. If the crop area is made easily accessible and easily adjusted, then this process could be easier.

In aquaponic there are two major things need to be taken care of – water flow and water quality. There are different water flow requirement for different type of aquaponic setup. Aquaponic most common components can be Growbed (mediabed), Nutrient Film Technique (NFT) or Deep Water Culture (DWC) [4].

C. Aquaponic Growbed

Common mediabed system component are cultivation tank, cultivation gravel (clay aggregate or pebbles) and siphon. There are three common area which are dry zone, root zone and sludge zone. Dry zone is normally 1 to 2 inches depth at the top area. Water definitely will not reach this area. The sludge area is the lowest area and considered all-time-wet area. The root area is the middle area which normally between 8 to 10 inches depth. Light Expanded Clay Aggregate (LECA) normally used as the grow media since it is near neutral pH and light in weight [8]. The granule feature of LECA or pebbles enables the air to exist between them.

The siphon functions to control the water level in the mediabed tank. It allows the water to fill the tank up to certain maximum level and then flush them out immediately. This gives the root of the crop to have contact with air. Hence, the result of this mechanical operation will permit the crop exposed to water and air alternately. Figure 2 shows the component of the growbed.

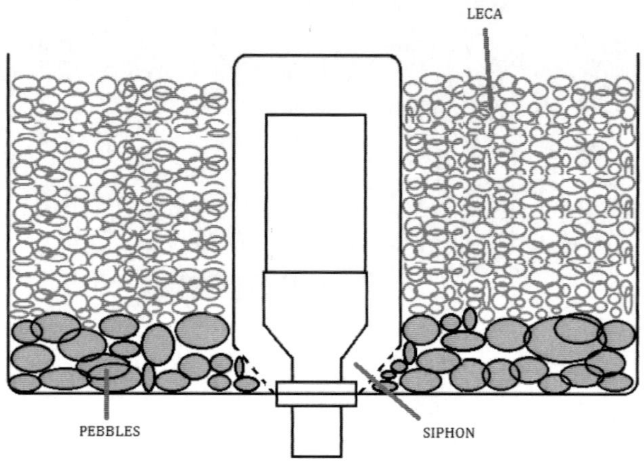

Figure 2: Growbed components

The flush operation improves the oxygen content in the filtered water. As stated by [9] the bell siphon should starts, drains, and stops every 15 – 20 minutes for optimal filtration and plant growth.

3. Test setup

A set of testing has been carried out using Affnan bell-strainer siphon design [10]. The siphon function and basic structure act like covered standpipe with a trapped outlet as described by [11]. The difference between Affnan siphon and the structure illustrated by Garret is at the end of the outlet pipe. Affnan's design just uses an elbow pipe and a straight pipe. However the latter create a crest height at the end of the pipe.

The siphon construction for this test and actual implementation will follow the one created by Affnan. This design was used since the construction need less effort and save time since the parts that were used available on the shelf. The guard area was constructed using 80mm unplasticized polyvinyl chloride (uPVC), covered by the end cap at the top and vent cowl at the bottom. Three siphon stand pipe sizes were prepared to be used in this test setup - 15mm, 20mm and 25mm PVC type. The vent cowl was drilled to house the suitable PVC tank connector. The base of the barrel also was drilled in order to install the siphon. At the bottom part PVC P/T connector fitting connected to the tank connector in order to tightly attach the siphon to the tank.

The test used normal tap water from main domestic outlet, feed to the liquid flow speed sensor and enter the half-cut barrel filled with LECA. Water inlet speed was controlled using mechanical ball valve. The initial setup set as in Figure 3. Flow speed sensor equipped both in the inlet and outlet section. The inlet water flow speed aimed to be varied from 100 up to 800 liter per hour (L/H).

Figure 3: Initial test setup

In order to implement fog to this system, a Raspberry Pi had been used as the broker, a personal computer (PC) as the client that subscribe the data and publish the control command. NodeMCU board used as the client that located at the growbed tank which subscribe to the PC command and publish the data.

The initial test involved the functionality of the siphon with the variation of inlet water speed. The inlet flow speed data together with the outlet flow speed data sent from the NodeMCU to the network. The data transmission will be monitored or supervised by the Raspberry Pi. The result tabled and plotted in the graph form to see how the siphon performs.

The setup later upgraded by equipping ultrasonic level sensor to the growbed. The ultrasonic sensor will report the current water level in the growbed. The PC will subscribe to this water level data together with inlet flow speed and outlet flow speed data.

4. Initial result

As previously stated, the ideal operation of a siphon at specified constant inlet flow speed is when the time between two flushes of water happens within 15 to 20 minutes. This timing range provides enough time for the bacteria in the growbed to extract nitrate and nitrite from ammonia. Bacteria involved in this process hence make this system the way a biofilter works.

Figure 4 shows the result of siphon operation at the inlet flow speed of 100 liter/hour for 15mm standard pipe size. As can be seen in the graph two flush and drain process has successfully happened. The flow in rate constantly exists along the process. The flow out rate was initially at zero rate and then spikes to the highest rate to flush out all the water in the barrel. The rate then goes to zero once the water level reaches to the lowest flushable level. The duration between two successive flushes for this inlet rate was around 18 minutes.

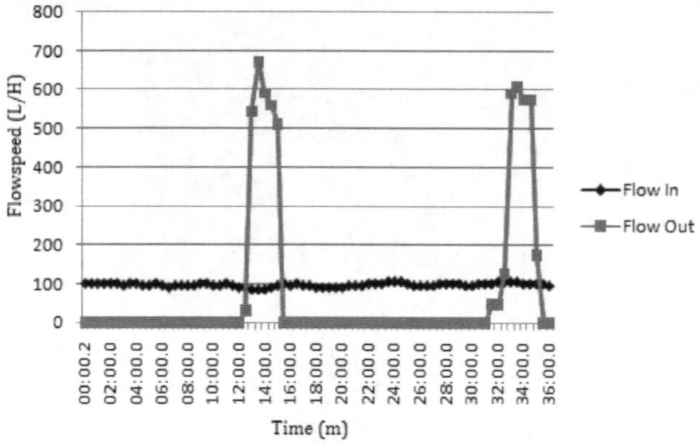

Figure 4: Flow pattern at 100L/H inlet flow speed

A problematic siphon happens when the cycle is too short or it cannot continuously fills and flushes alternately. Figure 5 shows a problematic siphon operation in term of timing. The test was done at the flow speed of 200 L/H. For this speed of water inlet, the duration between two successive flush was around 12 minutes.

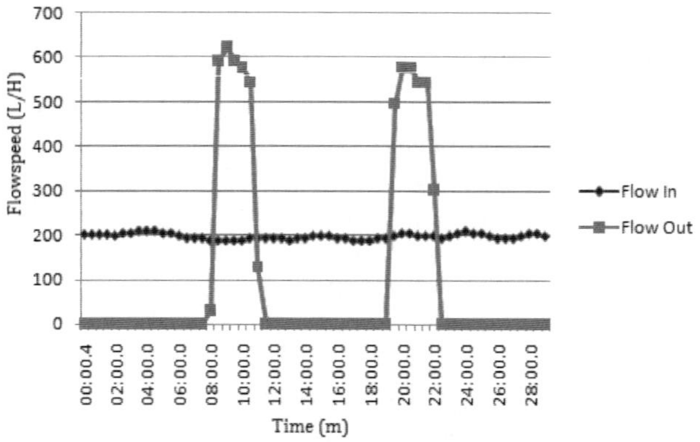

Figure 5: Flow pattern at 200L/H inlet flow speed

The following Figure 6 shows the flow pattern for 300 L/H water inlet for 15mm siphon standpipe size. This pattern exhibit a bit different shape compared to the previous flow pattern.

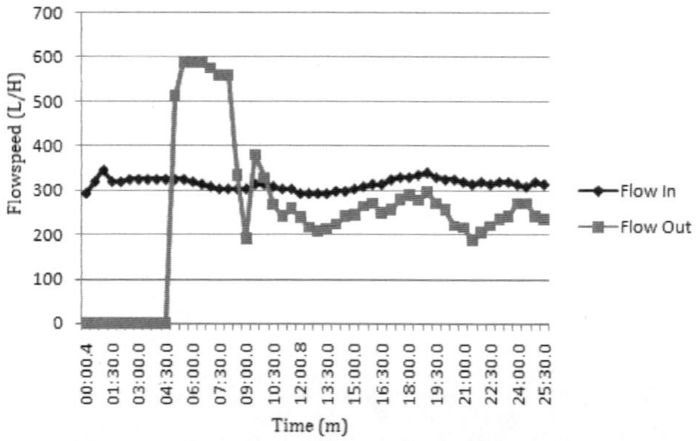

Figure 6: Flow pattern at 300L/H inlet flow speed

The same pattern also happen when the flow of 400 L/H feed to the growbed as shown in Figure 7. The flow out starts with the spike and then follows the inlet flow rate.

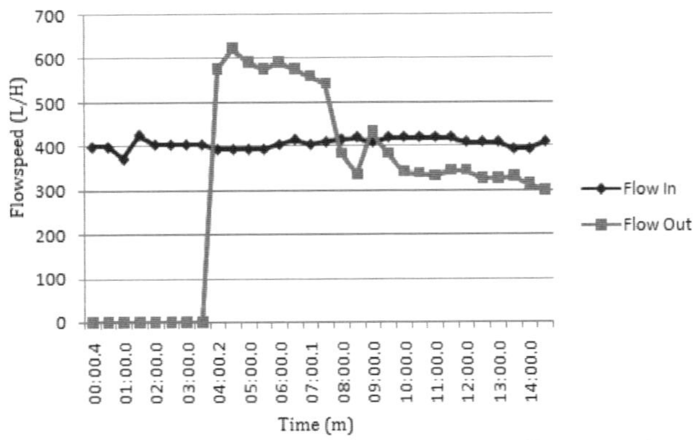

Figure 7: Flow pattern at 400L/H inlet flow speed

Both represent the flow pattern of malfunctioning siphon in term of flushing capability. Initially the siphon able to demonstrate the flushing operation but fail to repeat and this shown by the almost equal flow speed between the flow in and flow out. This means that the incoming flow will directly goes out at almost at the same rate.

Based on these flowspeed tests, it can be said that for 15mm standpipe siphon size placed at barrel-type growbed only 100 L/H inlet speed fit to the required operational condition. The siphon able to work well in term of flushing capability at 200 L/H inlet flowspeed but cannot obey 15 – 20 minutes fill-and-flush requirement. Meanwhile for 300 and 400 L/H inlet flowspeed obviously shows that the siphon cannot flush as desired. They can flush once only and then when the flow out rate equal to flow in rate they seems to be constant.

From these test results a fill-up table can be constructed. The table consists of the duration from the moment the flow-out rate become zero until the next flush out happened. Since the 100 L/H and 200 L/H inlet flowspeed show multiple fill-up duration, both of them will be recorded in Table 1 together with the average value.

Table 1 estimated fill-up time

Flow In Speed (L/H)	**Estimated Time**		
	Fill 1	**Fill 2**	**Average**
100	12m 28s	15m 56s	14m 12s
200	7m 58s	7m 38s	7m 48s
300	4m 50s	NA	4m 50s
400	3m 39s	NA	3m 39s

Before continuing to solve the problem by putting control elements, one more test has been done. This test involves water level checking. An ultrasonic sensor was placed in the siphon guard pipe and the water level kept published to the client. Figure 8 shows the water level graph when tested with 100 L/H flow speed for 15mm siphon standpipe.

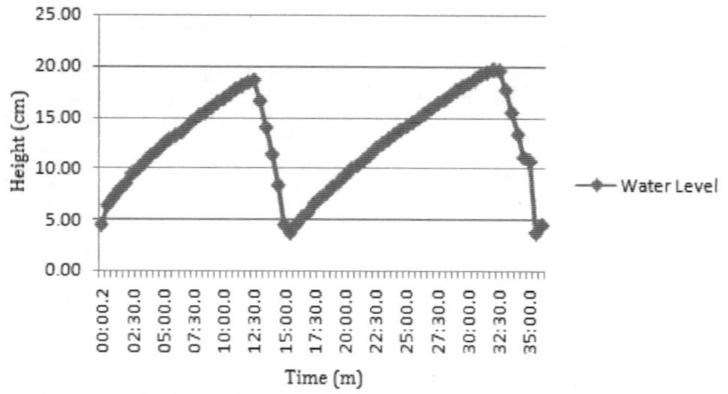

Figure 8: Level pattern at 100L/H inlet flow speed

The height data represent the real water level in the growbed. The values calculated by substracting the sensor position with the read value. The difference will give the actual water level in the siphon.

From this graph the lowest level the water can go is around 4 cm. This can be said as the sludge area height. From the graph also we can see that the maximum height achievable was around 19.5cm. This was the highest level can be reached before the flush process happen.

5. Implementation of control

Siphoning mechanism is not only good to the crop by providing the nutrient but also beneficial to fish since it helps to oxygenate the flow out water. What can be done is to improve the design by removing the flow speed dependency, controlling the flushing duration while maintaining the good aspect of siphon.

From both Figure 4 and 5, the duration between the start of flush until the end of it can be estimated constantly at 3 minutes for this 15mm siphon pipe size. Meanwhile from Figure 8 the maximum water level before flush out is around 19.5cm.

In order to make the siphon able to function well regardless of the flow in speed, adjustment needs to be done to the flow in part. The suggested adjustment was by stopping the flow in before the water level reach to flush level. Figure 9 shows the water level pattern introduced for 100 L/H flow speed.

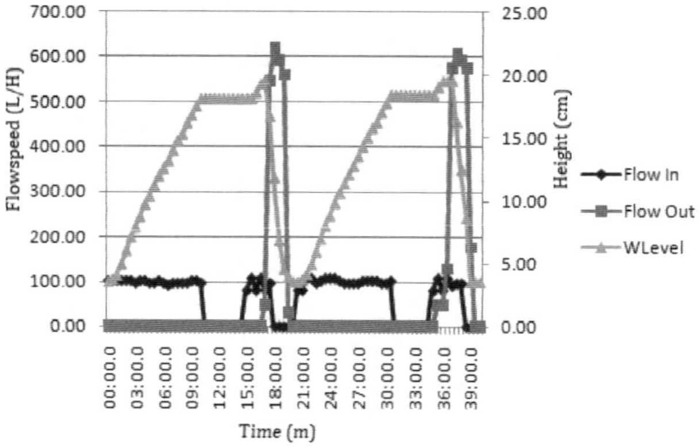

Figure 9: Flow pattern at 100 L/H with water level detection

The following Figure 10 shows the level pattern for 200 L/H inlet speed. The flow in cut-off time come around the 6th minute and it needs 9 minutes to wait before further fill up can be added to force flushing condition.

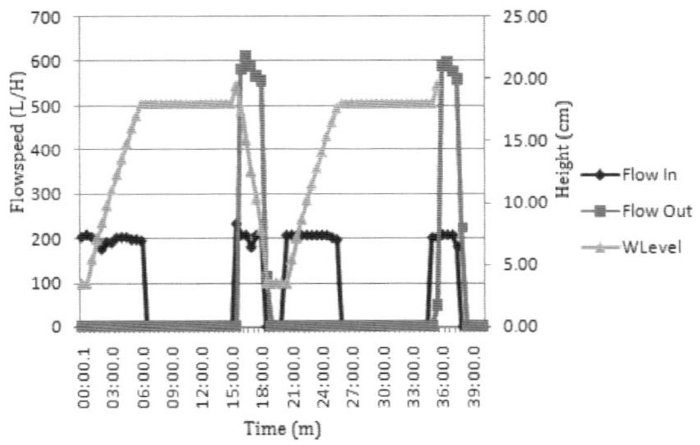

Figure 10: Flow pattern at 200 L/H with water level detection

The following Figure 11 shows the level pattern for 300 L/H inlet speed. The cut-off time come around the 3rd minute and it needs 12 minutes to wait before further fill up can be added to force flushing condition.

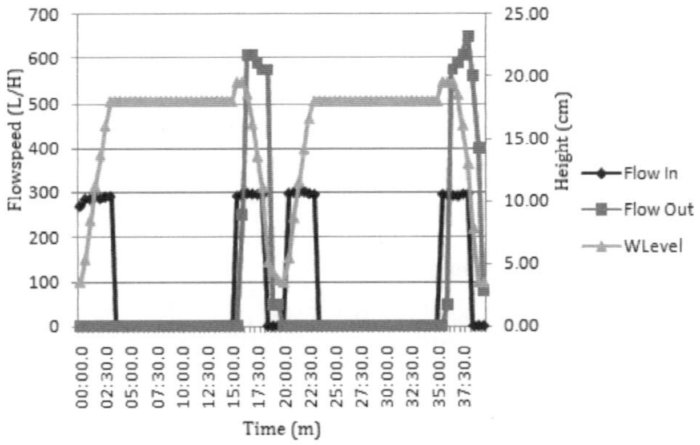

Figure 11: Flow pattern at 300 L/H with water level detection

In this water level control adjustment experiment the level before the flow in cut off was set to 18 cm. The wait time then introduced until the 15th minute of the cycle. The flow in then re-introduced for around 3 minutes to force the flow out. The water level experiment for 400 L/H flow in was also exhibit the same pattern. The set level reached at faster rate and the wait time needed to be longer.

6. Conclusion

Growbed replaces the need for costly biofilter in recirculating aquaculture system. By performing as a trickling filter this growbed together with beneficial bacteria helps to remove nitrate and nitrite from the fish water. Hence the catastrophic ammonia level can be reduced.

Preliminary test shows that mechanical setup can lead to unsatisfactory operation when operated with various flow-in rates. The unsatisfaction happens once the fill-and-flush operation fails to be in sequence or in term of timing.

From the mechanical operational analysis also the flow pattern can be examined and solutions can be planned for this problem. The solution plays around with the flow-in control.

The first solution can be done by controlling the flow-in based on the accumulated volume. The flow-in can be stopped before the total accumulated volume reach to 24.6 L as calculated before. This way will need to have flow sensor at the inlet channel.

The second solution can be done by controlling the flow-in based on the flow timing. Each flowrate should have pre-calculated duration of valve on and off. By doing this way also there will be a need to have flow sensor.

The third solution which is based on the water level seems to be the best since it involve simpler method and cheaper component.

Acknowledgement
Thank you to everybody that contributes to this research setup, analysis and writing. This valuable help really important towards the preparation of complete IoT based distributed system in aquaponics.

References

[1] Hanes D., Salgueiro G., Grosetete P., Barton R., Henry J., IoT Fundamentals: Networking Technologies, Protocols, and Use Cases for the Internet of Things, Cisco Press, Indiana, 2017
[2] Sinclair B., IoT Inc. How Your Company Can Use the Internet of Things to Win in the Outcome Economy, McGraw Hill Education, 2017
[3] Klein S., IoT Solutions in Microsoft's Azure IoT Suite: Data Acquisition and Analysis in the Real World, 1st. Ed., Apress, Washington, 2017
[4] *Somerville C., Cohen M., Pantanella E., Stankus A. and Lovatelli A.*, Small-scale aquaponic food production – Integrated Fish and Plant Farming, FAO Fisheries and Aquaculture Technnical Paper 589, Food and Agriculture Organization of United Nations, Rome, 2014
[5] Hillar G. C., MQTT Essentials – A Lightweight IoT Protocol, Packt Publishing Limited, United Kingdom, 2017
[6] https://blogs.cisco.com/digital/fog-computing-bringing-cloud-capabilities-down-to-earth?dtid=osscdc000283
[7] Bernstein S., Aquaponic Gardening – A Step by Step Guide to Raising Vegetables and Fish Together, New Society Publishers, Canada, 2011
[8] https://en.wikipedia.org/wiki/Expanded_clay_aggregate
[9] Fox, B. K., Howerton, R., and Tamaru, C. S., Construction of Automatic Bell Siphons for Backyard Aquaponic Systems, Biotechnology, College of Tropical Agricultural and Human Resources, 2010
[10] http://www.affnanaquaponics.com/2014/04/strainer-bell-siphon-simplified-affnan.html
[11] Garret, R. E., Principles of Siphons, Journal of the World Aquaculture Society, Vol. 22, No. 1, 1991

IOP Conf. Series: Journal of Physics: Conf. Series **1018** (2018) 012015 doi:10.1088/1742-6596/1018/1/012015

Performances of Machine Learning Algorithms for Binary Classification of Network Anomaly Detection System

[1]Mukrimah Nawir, [1]Amiza Amir, [1]Ong Bi Lynn, [1]Naimah Yaakob, [2]R.Badlishah Ahmad

[1]Embedded, Network, and Advanced Computing Research Cluster (ENAC)
School of Computer and Communication Engineering, Universiti Malaysia Perlis (UniMAP), Pauh, Perlis
[2]Faculty of Informatics and Computing, Universiti Sultan Zainal Abidin (UniSZA), 22200, Besut, Terengganu

E-mail: mukrimah1992@gmail.com; amizaamir@unimap.edu.my; bilynn@unimap.edu.my; naimahyaakob@unimap.edu.my; badli@unisza.edu.my

Abstract. The rapid growth of technologies might endanger them to various network attacks due to the nature of data which are frequently exchange their data through Internet and large-scale data that need to be handle. Moreover, network anomaly detection using machine learning faced difficulty when dealing the involvement of dataset where the number of labelled network dataset is very few in public and this caused many researchers keep used the most commonly network dataset (KDDCup99) which is not relevant to employ the machine learning (ML) algorithms for a classification. Several issues regarding these available labelled network datasets are discussed in this paper. The aim of this paper to build a network anomaly detection system using machine learning algorithms that are efficient, effective and fast processing. The finding showed that AODE algorithm is performed well in term of accuracy and processing time for binary classification towards UNSW-NB15 dataset.

1. Introduction

Security-based anomalies are an abnormality or malicious behaviours where the network or systems deviate from their realm usual functionality. This wide-ranging issue in network security need defences or protection tools either software or hardware-based tools that ensure the system in a full-secure from compromise by the attackers. For instance, during the communication or transferring data information between users they might compromise by hackers that have intention to stealth the information. Machine Learning (ML) is one of the efficient and modern technique that enable to monitor the patterns in highly system environment that present today [1].

However, network anomaly detection system using machine learning faced difficulty when dealing the involvement of dataset. There are several issues regarding network anomaly detection dataset for evaluation purpose with the present dataset as discussed in paper [2]. Most datasets not fulfilled the requirements for network security. The labelled dataset is very few in public [3] and the conflicts arise when some of this dataset only specific for certain environment (flexibility), less availability, and a lack of ground truth as state in [4].

Hence, in this paper we construct the network anomaly detection systems that capable to manage the large-scale data and frequently exchange their data through Internet by employing the machine learning (ML) algorithms onto it. The challenges and issues of network anomaly detection datasets motivate this paper to investigate the UNSW-NB15 dataset. The creation of this dataset consists a hybrid of real normal and synthetic network attack data as presented in [5]. Another point is that, UNSW-NB15 dataset overcome the problems faced by researchers with the others dataset where they represent the complex patterns that have balanced data distribution of training and testing set. Advantageous of UNSW-NB15 dataset reflect the high efficiency due to the features of instances collect from the payload to the header of information [6].

The paper organized as followed: In Section 2 present the existing work with the current issues of labelled network dataset. Next, the experimental set-up describes details at Section 3. From the conducted experiments, Section 4 discussed the finding of experiments. Last part of this paper to conclude (Section 5).

2. Related Works

Numerous number of network anomaly detection dataset announced by many community such as NSL-KDD (improvement of KDDCup99 dataset), MAWILab [7], the MoMe Cluster, the Cooperative Association for Internet Data Analysis (CAIDA) and RIPE comprehensively critics by many researchers. For instance, CAIDAs datasets not effective to be used for network anomaly detection where their data resources might be removed during the simulation and it is not adaptable to any environment.

Several works had been done for classification to determine the behavior of data in a system. Authors [8] compared the UNSW-NB15 and KDDCUP99 dataset to measure the accuracy and False Alarm Rate (FAR) using five different ML algorithms and found that DT is an efficient algorithm for classification differ from our paper that investigate the accuracy and other measures by binary classes (Normal (1) and Attack (0)).

In addition, the most famous dataset that had been used by many researchers, DARPA datasets (KDDCup1999), are perceived to skewness and biases classification toward the training set that repetitious their data instances [5][9][10]. Hence, these dataset is impractical for network anomaly detection systems. Moreover, since the data is dynamic in the system causes the new attack might be present in the network. These old-fashioned network anomaly detection dataset also in-comprehensive to represent a modern normal behaviours and contemporary synthesized attack as in UNSW-NB15 dataset [11].

The latest related work is present in paper [12] stated that the datasets consists irrelevant and redundant attributes. They believe that by using features selection result a fast processing and high in accuracy of classifier. They employed ML techniques (Random Forest) toward the datasets. From here motivate this present paper to observe the performance of time taken to train and test the model with varying the size of data.

3. Experiment Setting

In our experiment, Ubuntu software version 13.10-0 ubuntu 4.1 is the operating system and the WEKA tool [13] is run on an Intel Xeon (R) CPU E3-1270 v5 @ 3.60GHz x 8, 16GB RAM to employ ML algorithms toward UNSW-NB15 dataset to evaluate based on their classification rate and processing time in order to build a high efficient and fast processing for network anomaly detection system.

According to Fig. 1 the binary classification for network anomaly detection system toward UNSW-NB15 dataset involved four stages (preparation of dataset, training and testing, build classifier model and performance measure metrics). The experiment begins by loading a network dataset that needed for classification purpose. Once the dataset is ready, the data will be undergoing the training/testing stage. Before testing stage, the classifier model is required as a decision engine. Finally, to analyse the result from the conducted experiments.

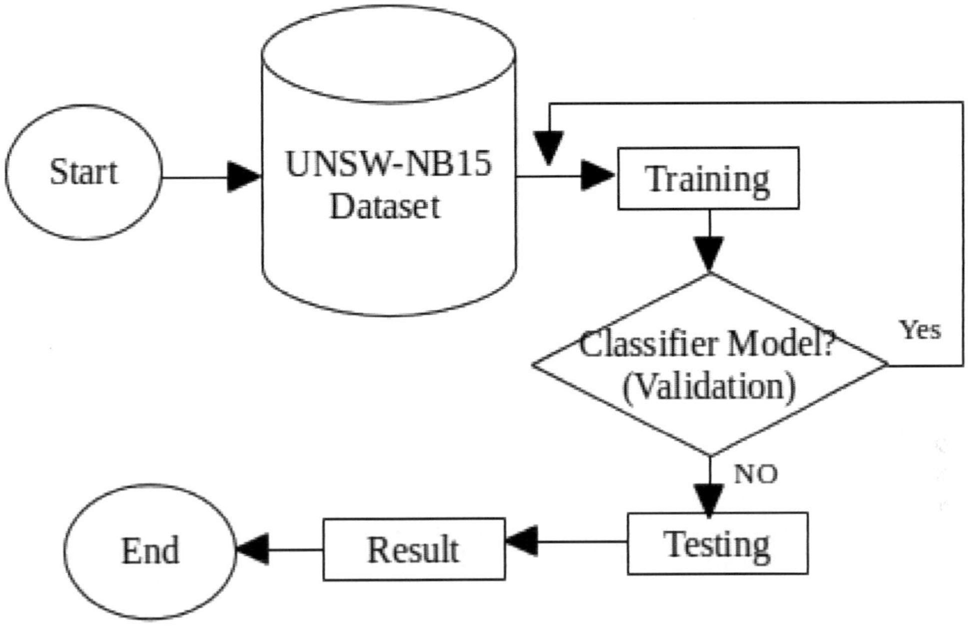

Figure 1. Binary Classification (UNSW-NB15 dataset) using Machine Learning Techniques

3.1. Labelled Network Dataset – UNSW-NB15

Presently, the labelled network dataset that used for network anomaly detection is UNSW-NB15 dataset. In the papers [14][15], describe details on this dataset. But, in this paper we make some modification toward the original UNSW-NB15 dataset where the number of attributes (features) only 43 features instead of 45 features. It is important to select certain features to get the reliability of the experiment. Feature (*id*) excluded due to it is just an index and a row of numbers. If it is not excluded the experiment is not reliable because they affect the behaviour of the whole dataset.

Since, the UNSW-NB15 dataset provide two labelled features (i.e. *cat-attack* and *label*). We only used label attributes where the data is either normal or attacks data (binary data). Otherwise, this dataset also can be done for a multi-classification by choosing *cat-attack* as label features instead of *label*. These 43 features/attributes according to their type of features (such as basic, flow, time, content, connection, general, and label) used for this present work is as tabulated in Table 1.

IOP Conf. Series: Journal of Physics: Conf. Series **1018** (2018) 012015 doi:10.1088/1742-6596/1018/1/012015

Table 1. Features of UNSW-NB15 Dataset

No.	Name	Features	No.	Name	Features
1	*dur*	Basic	23	*dwin*	Content
2	*Proto*	Flow	24	*tcprtt*	Time
3	*Service*	Basic	25	*synack*	Time
4	*State*	Basic	26	*ackdat*	Time
5	*spkts*	Basic	27	*smean*	Content
6	*dpkts*	Basic	28	*dmean*	Content
7	*sbytes*	Basic	29	*trans-depth*	Content
8	*dbytes*	Basic	30	*response_body_len*	Content
9	*Rate*	Basic	31	*ct_srv_src*	Connection
10	*sttl*	Basic	32	*ct_state_ttl*	General
11	*dttl*	Basic	33	*ct_Dst_ltm*	Connection
12	*sload*	Basic	34	*ct_Src_Dport_ltm*	Connection
13	*dload*	Basic	35	*ct_dst_sport_ltm*	Connection
14	*sloss*	Basic	36	*ct_dst_Src_ltm*	Connection
15	*dloss*	Basic	37	*is_ftp_login*	General
16	*sinpkt*	Time	38	*ct_ftp_cmd*	General
17	*dinpkt*	Time	39	*ct_flw_http_mthd*	General
18	*sjit*	Time	40	*ct_src_ltm*	Connection
19	*djit*	Time	41	*ct_srv_dst*	Connection
20	*swin*	Content	42	*is_sm_ips_ports*	General
21	*stcpb*	Content	43	*label*	Labelled
22	*dtcpb*	Content			

3.2. Machine Learning Tasks

Generally, machine learning tasks involved three schemes (Training, Validation and Testing). Firstly, the dataset is train the classifiers for binary classification (either normal or attack data). During training stage, the features/attributes (either numeric or nominal) likewise source and destination network prefixes, time slot, and etc. are extracted and known to representing the patterns of data. Next, we test our training model using tenfold cross validation to start the evaluation of the classification model for network anomaly detection.

3.3. Build Classifier Model

We employed three types of ML algorithms from Bayesian's family in WEKA tools. They are Average One Dependence Estimator (AODE), Bayesian Network (BN), and Naive Bayes (NB) to determine the performance in term of classification rate and processing time for classifier model to classify the data instances of UNSW-NB15 dataset. The parameters of these classifiers set to default as in WEKA and using tenfold cross validation to validate the training set before the model been tested.

3.4. Performance Measure

In the context of ML, metrics measure such as accuracy, true positive rate (TP Rate), False Positive Rate (FP Rate), Precision and Recall that representing the performance of binary classification. In addition, this paper presents the varying size of training data with 10 samples (10 000, 20 000, 30 000, 40 000, 50 000, 60 000, 70 000, 80 000, 90 000, and 100 000 data instances) affects the time taken to train and test the classifier model.

IOP Conf. Series: Journal of Physics: Conf. Series **1018** (2018) 012015 doi:10.1088/1742-6596/1018/1/012015

4. Result and Discussion

4.1. Performance of ML algorithms for Network Anomaly Detection System

The discussion of the results begins with comparing the AODE algorithms with Bayesian Network and Naive Bayes in term of the correctly classification the data instances (accuracy) and time taken to model the classifier. It is apparent from the chart that (Figure 2) AODE is efficient as well as fast processing for network anomaly detection system compared to another two classifiers with the percentage of accuracy equal to 94.37% and only need to train the model in 4.13s. Meanwhile, BN algorithm that performance almost closed to AODE algorithm with the accuracy is 92.70% and time taken is 4.17s.

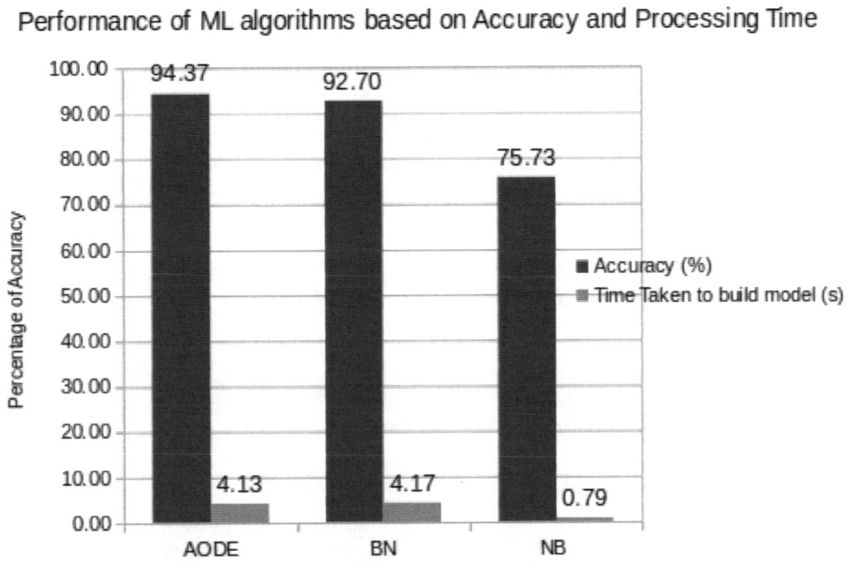

Figure 2. Performance of ML Algorithms for Network Anomaly Detection System

Although, the Naive Bayes algorithm required small amount of time (with 0.79s) to classify the data instances of given dataset (UNSW-NB15) but it is not comparable to AODE and BN algorithms based on classification rate. To measure the performance of this three ML algorithms not enough to consider only their accuracy, the performances of metrics measure (i.e. True Positive Rate(TP Rate), False Positive Rate (FP Rate), Precision (Prec), Recall) of ML algorithms for these binary data (Class 1=Normal and Class 0=Attacks) using UNSW-NB15 dataset need to be investigate as well that tabulated in Table 2.

Table 2. Performance of Supervised ML algorithms for Network Anomaly Detection System

ML Algorithm	TP Rate		FP Rate		Prec		Recall	
	Class 1	Class 0	Class 1	Class 0	Class 1	Class 0	Class 1	Class 0
AODE	0.941	0.95	0.050	0.059	0.976	0.882	0.941	0.95
BN	0.957	0.863	0.137	0.043	0.937	0.904	0.957	0.863
NB	0.675	0.932	0.068	0.325	0.955	0.574	0.675	0.932

4.2. Effect of Time Taken against the Varying Size of Training Data

In our concern, the time taken to build classifier model is enumerating to have an efficient and effective with a fast classification for network anomaly detection system. Figure 3 shows that the effect of time taken by varying the number of training data. The square point is for time taken to build the classifier whereas the cross-point result of time taken to test the model.

The finding showed that NB algorithm is fastest to build the model for network anomaly detection system of binary classification during training or testing scheme where the time taken with increasing the size of training data is in the range of 0.77s-0.94s to train them and to test the model required 1.03s to 2.16s only. As aforementioned that the accuracy of this algorithm is lowest compared the other two ML algorithms. Moreover, the NB algorithm become faster to test the model when the number of training data increases.

Even though, to train the AODE algorithm take a long time compared to NB, they are more fast to test the built model and the correctness to classify the data instances is more robust. Yet, this training time of AODE is linearly with respect to the size of training data and can be learn in incremental way. For instance, with the 50k training data the differences time taken for training and testing for these both algorithm where AODE need double amount of time compared NB algorithm. BN algorithm's performances, when the large number training data caused the low time taken to train as well as to test their model. This classifier considers slower during training process.

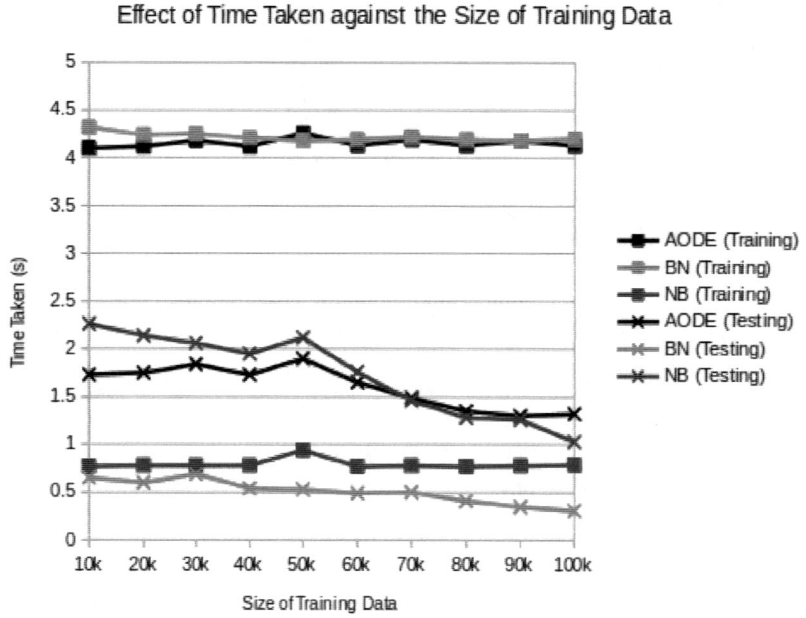

Figure 3. Effect of Time Taken against the Size of Training Data

5. Conclusion

As a conclusion, the UNSW-NB15 dataset is a public network dataset that is relevant for network anomaly detection due to their patterns are complex that represented in modern as well as in contemporary synthesized attacks. From the evaluation made in this paper using ML algorithms (Bayesian group) found that the AODE algorithm is an outperformed, effective, efficient, and fastest classifier for network anomaly detection of binary classification toward the network labelled (UNSW-NB15) dataset.

Acknowledgments
The research reported in this paper is supported by Research Acculturation Grant Scheme (RAGS). The authors would also like to express gratitude to the Malaysian Ministry of Higher Education
(MOHE) and University Malaysia Perlis for the facilities provided.

References
[1] B.D.Kumar, K.J. Kumar, Network anomaly detection: A machine learning, CRC Press, 2013.
[2] S. Ali, S. Hadi, T. Mahbod, G.A.Ali, Toward developing a systematic approach to generate benchmark datasets for intrusion detection, computers & security, pp. 357{374, vol. 31, no.3, 2012, Elsevier.
[3] G.Folino, F.S Pisani, P.Sabatino, A Distributed Intrusion Detection Framework Based on Evolved Specialized Ensembles of Classifiers,Applications of Evolutionary Computation: 19th European Conference, EvoApplications 2016, Porto, Portugal, March 30 { April 1, 2016, Proceedings, Part I", Springer International Publishing, pp.315-331, 2016.
[4] R. Koch, M. Golling, G.D. Rodosek, 11 - Towards Comparability of Intrusion Detection Systems: New Data Sets, TNC2014 - TNC2014. (n.d.). Retrieved August 11, 2017, from https://tnc2014.terena.org/core/poster/13.
[5] N. Moustafa, J. Slay, UNSW-NB15: a comprehensive data set for network intrusion detection systems (UNSW-NB15 network data set). 2015 Military Communications and Information Systems Conference (MilCIS), pp. 16, 2015. https://doi.org/10.1109/MilCIS.2015.7348942.
[6] N. Moustafa, J. Slay, A hybrid feature selection for network intrusion detection systems: Central points A HYBRID FEATURE SELECTION FOR NETWORK INTRUSION DETECTION SYSTEMS: CENTRAL POINTS AND ASSOCIATION RULES, pp. 513 2015.
[7] F. Romain, B. Pierre, A. Patrice, F. Kensuke, MAWILab: Combining Diverse Anomaly Detectors for Automated Anomaly Labeling and Performance Benchmarking, ACM CoNEXT '10, p. 12, December,2010.
[8] N. Moustafa, J. Slay, The evaluation of Network Anomaly Detection Systems: Statistical analysis of the UNSW-NB15 data set and the comparison with the KDD99 data set, Information Security Journal: A Global Perspective, vol. 25, pp.18-31, 2016.
[9] F. Gumus, C. Sakar, Z. Erdem, O. Kursun, Online Naive Bayes classification for network intrusion detection, ASONAM 2014 - Proceedings of the 2014 IEEE/ACM International Conference on Advances in Social Networks Analysis and Mining, (Asonam), pp.670-674, 2014.
[10] O.Osanaiye, H.Cai, K.K.R Choo, A.Dehghantanha, Z. Xu, M. Dlodlo, Ensemble-based multi filter feature selection method for DDoS detection in cloud computing, EURASIP Journal on Wireless Communications and Networking, pp.130-139,2016.
[11] D. G. Mogal, S. R. Ghungrad, B. B. Bhusare, A Review on High Ranked Features based NIDS. Ijarcce, 6(3), pp. 349353 2017. https://doi.org/10.17148/IJARCCE.2017.6380.
[12] J.Tharmin, Z. Shahrzad, Feature selection in UNSW-NB15 and KDDCUP'99 datasets, Industrial Electronics (ISIE), 2017 IEEE 26th International Symposium on, pp.1881-1886, IEEE, 2017.
[13] Weka 3-Data Mining with Open Source Machine Learning in Java, (n.d), Retrieved July 11, 2017, from http://www.cs.waikato.ac.nz/ml/weka/.
[14] The UNSW-NB15 data set description, https://www.unsw.adfa.edu.au/australian-centre-forcybersecurity/cybersecurity/ADFA-NB15-Datasets/, 2 March, 2016, Accessed date: sept,2017.
[15] N. Moustafa, J. Slay,The UNSW-NB15 data set description, 2 March 2016. https://www.unsw.adfa.edu.au/australian-centre-for-cyber-security/cybersecurity/ADFA-NB15-Datasets/.

IOP Conf. Series: Journal of Physics: Conf. Series **1018** (2018) 012016 doi:10.1088/1742-6596/1018/1/012016

Initiating an Online Reputation Monitoring System with Open Source Analytics Tools

Mohd Ilias M Shuhud, Najwa Hayaati Md Alwi and Azni Haslizan Abd Halim

Faculty of Science and Technology, University Sains Islam Malaysia, Nilai, Malaysia

ilias@usim.edu.my

Abstract. Online reputation is an invaluable asset for modern organizations as it can help in business performance especially in sales and profit. However, if we are not aware of our reputation, it is difficult to maintain it. Thus, social media analytics is a new tool that can provide online reputation monitoring in various ways such as sentiment analysis. As a result, numerous large-scale organizations have implemented Online Reputation Monitoring (ORM) systems. However, this solution is not supposed to be exclusively for high-income organizations, as many organizations regardless sizes and types are now online. This research attempts to propose an affordable and reliable ORM system using combination of open source analytics tools for both novice practitioners and academicians. We also evaluate its prediction accuracy and we discovered that the system provides acceptable predictions (sixty percent accuracy) and demonstrate a tally prediction of major polarity by human annotation. The proposed system can help in supporting business decisions with flexible monitoring strategies especially for organization that want to initiate and administrate ORM themselves at low cost.

1. Introduction

In early October, 2017, the government of a Middle-Eastern country banned four Malaysian universities for administrative issues [1]. The incident raised various reactions, especially by local and international netizens, which indirectly or directly affected the reputation of the involved universities. Furthermore, there were rumours made on social media which claimed that the incident was primarily due to education quality. However, in another report, the vice chancellor of one of the universities took a quick action to refute the rumour with the fact that the reason involved administrative matters [2]. It is expected that the management of each of the affected universities will strategize for rectification of the damage caused to its reputation and public image. However, they might have interested to the current status of their online reputation.

The above scenario is an example of an online reputation attack, which is a growing concern of many organizations today that is probably more harmful than computer viruses. The rapid development of social media has challenged various aspects of a business, including online reputation. While positive reputation can influence purchases, investment and loyalty towards products or services, a negative reputation may have an opposing effect, e.g., customer attrition, sales reduction, and distrust. Therefore, organizations should be aware of their reputation by monitoring what people say about our brands or products. “If a company is unaware of its own reputation, it cannot manage it successfully” [3]. One way to manage reputation is by implementing online reputation management to support business decisions [4], [5]. Unfortunately, many organizations do not implement ORM due to cost, complexity and reliability [6]. For example, the cost of negative review removal in search engine can be from RM50,000 (USD 12,800) according to an ORM company.

This research attempts to propose an affordable and reliable ORM system using open source data analytics tools and then evaluate its performance. The research contributes in enabling an online reputation monitoring implementation for any type and size of organizations at almost no cost. It also allows academic research to explore social media data for various purposes. In this research we combine, different open source tools such as data science software platforms, business intelligence platforms, text analytics Application Programming Interfaces (API) and Microsoft Excel Online. We began with a background study of the existing literature, followed by a context study which involved an interview. Later, we conducted experiments to evaluate the developed ORM system environment. We discovered that the system produced acceptable polarity prediction accuracy (over 50 percent of human annotation) and produced similar prediction of major polarity type with human annotation in sentiment analysis of a news.

This paper is organized as follows. The next section reviews related work to provide the background of important keywords in the research and theoretically support the study. Next, we present the research methodology in Section 3, and then present our experiment results and evaluate selected analytics tools in Section 4. In Section 5, we provide some insights from the experiments, including advantages and limitations. Finally, we conclude in Section 6.

2. Background

This section provides a background on the key terms in the research from the relevant literature.

2.1. Social media analytics

In defining social media analytics, [7] describes it as "the practice of gathering data from social media websites and analysing the data using social media analytics tools to make business decisions". The definition highlights making business decisions as the ultimate goal, regardless of the type of organization whether commercial, government and non-profit organizations. The definition by [7] is in-lined in a more comprehensive definition by [8] in his book, where "social media analytics is the art and science of extracting valuable hidden insights from vast amounts of semi-structured and unstructured social media data to enable informed and insightful decision making. It is a science, as it involves systematically identifying, extracting, and analysing social media data (e.g., tweets, shares, likes and hyperlinks) using sophisticated tools and techniques. It is also an art, interpreting and aligning the insights gained with business goals and objectives. To obtain value from analytics, one should master both its art and science"[8].

Social media analytics falls under interdisciplinary fields in computer sciences and social sciences. The practice of social media analytics falls under computer science, however, the analysis produced is useful in a wide spectrum of disciplines in social science such as political science, sociology, geology, business and marketing and journalism [9]. Among the various purposes of social media analytics in commercial organizations is to support marketing and customer service processes by mining customers' sentiment from comments, review, and tweets. In social media analytics, the most critical step is to determine how the collected and analysed social media data can benefit specific business goals [8].

2.2. Online reputation monitoring

Online reputation monitoring or management (ORM) is a new element of public image for entities, especially organizations that establish their presence online. Online reputation can be defined by search results on a search engine, social media comments, a ranking of consumer reviews, articles and news about the organization or the related attributes such as products and services [10]. What do we see when we search for our name and company name on Google or Facebook? If the first page is all good about us, then this is a sign of good online reputation. But if it is bad information, we might be in trouble, as people normally do not check the next pages. Large organizations such as eBay.com and Amazon.com have embedded an ORM in their e-commerce platform so that customers can research how people think about and rate certain products. This is because 90 percent of online purchases are driven by product reviews [11]. This shows that online reputation is an invaluable intangible asset that should be monitored by organization of all sizes, types, and popularities. Some companies or government departments outsource the ORM to a third-party company for analytics services, while others employ data analysts to work internally [12]. Among of the useful tools in ORM is sentiment analysis, which provides an overall binary polarity value for an entity's online reputation. There are many works in the literature that have discussed ORM in various industries such as healthcare, politics, journalism, hotels, tourism and automotive. For example, in [13], researchers suggest that online reputation with sentiment analysis can enhance competitiveness in the automotive industry.

2.3. Sentiment analysis

Sentiment analysis has been established as among the hot research topics across domains in the past decade [14]. It is one of many content analysis methods in social media analytics, and is also called opinion mining, subjectivity analysis, and appraisal extraction. Sentiment analysis is essentially a natural language processing task that involves mining attitudes, opinions and emotion from text or speech [15]. Sentiment analysis is a classification problem where opinionated texts are categorized as "positive", "negative" or "neutral" using either machine learning algorithms or predefined lexicons [16]. Businesses can use sentiment analysis to understand consumers' emotional reactions towards their brand, company or signature products and services, and whether they are accepted positively or negatively on the Internet [17].

2.4. Related works on social media analytics using open source tools

There are currently many ORM companies offering their solutions. However, the service cost varies based on the needs and difficulty [6]. Thus, we attempt to explore other options in open source systems. In the work of [18], researchers attempt to develop a Java program to fetch real-time data from various social media platforms through various APIs for analysis purposes. Although they managed to extract some data, there are some restrictions set by social media platforms that limited the output. Meanwhile, in [19], an open source social analytics tool called Netlytic was evaluated. The researchers gathered contents from Facebook and YouTube for analytics and proved that social media analytics is not costly if we are able to explore alternative ways rather than to purchase predefined software. In the work of [20], the researchers discusses the ethical and legal issues on gathering social media data from APIs, which is another important angle that should not be overlooked.

Besides social media data gathering, [21] introduced an open source API for text analytics, crawling and interpretation called TACIT. The API features an intuitive tool and interface to utilise the technologies in text analysis and big data management for use by other researchers. Using multiple APIs in developing analytics environment can be a complex process. [22] provides a relevant method mapping across APIs called TMAP by using text mining on natural language API method descriptions. The method can assist software developers to migrate applications from a source API to the target API automatically.

3. Research methodology

In this research, we divided the research methodology into three phases, namely: (1) Foundation Phase (2), Experimentation Phase, and (3) Evaluation Phase.

3.1. Foundation phase

In this phase, we have conducted three sub-phases of groundwork, which are; background study, context study and tools selection.

3.1.1. Background study

We have conducted a review of relevant studies in the literature around important key words such as online reputation, online reputation management, social media analytics, user-generated contents (UGC) and sentiment analysis. The online databases used for searching for the literature are Google Scholar, Scopus, Mendeley and others. Part of the review is presented in Section 2.

3.1.2. Context study

To acquire a deeper understanding on the context of the study, especially online reputation monitoring practices, we have conducted an interview with a key person at a public relations unit of one of the involved universities in the banned incident mentioned above. The collaboration with the public relation unit of the university has been started a week before the banned issue was raised. The interview was transcribed and analysed for a better understanding on the practices by the organization.

3.1.3. Tools selection

Furthermore, we select open source data analytics tools which are appropriate for the experiments. The tools are required for social media data extraction, text processing, sentiment analysis, and report analysis. Three aspects are considered in selecting the appropriate tools: ease of use, accessibility, and reliability. Ease of use means that the tools are not too complex for a non-IT background user and a graphical user interface (GUI) is necessary. Accessible criteria mean that we have the access to the tools in terms of knowledge, e.g., training and acquisition of the tools. In terms of reliability, this refers to open-source text analytics APIs, which include sentiment analysis module in the package. For this, we need a tool that provides acceptable accuracy, or at least over fifty percent from human annotated polarity prediction. As mentioned earlier, we have selected a data science software platform called RapidMiner Studio for data processing which allows 10,000 rows per transaction to be processed in the free version. For sentiment analysis, we used a machine learning text analytics API which will be referred as 'M API' in this paper. M API is among of text analytics API available in RapidMiner Marketplace. In extracting comments from Facebook, we used Microsoft Power BI, an open source business intelligence and analytics program. We chose Facebook comment as it has more than 140 characters in a single post, unlike Twitter. Thus, more texts can be analysed. As for customization and filtration of the extracted data and results, we used Microsoft Excel Online a free version of the software included in Microsoft Office Online or Microsoft Office 365.

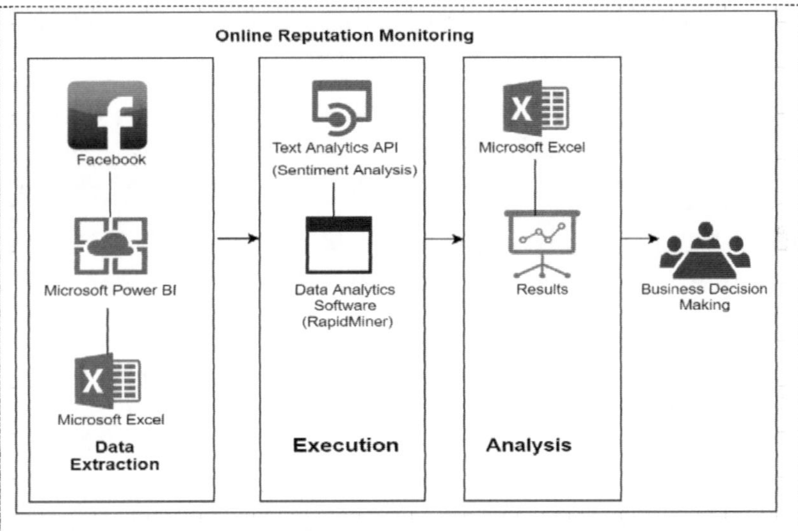

Figure 1. Main processes involved in the Online Reputation Monitoring System using opensource tools

3.1.4. Experimentation phase

In this phase, we conducted two experiments to evaluate the accuracy of the selected API in predicting sentiment polarity of UGC and to evaluate the system capability to perform sentiment analysis towards bad or sensitive news related to an organization.

3.1.5. Evaluation phase

In the evaluation phase, we provide some insights on the performance of the open source analytics tools in the experimentation phase. In summary, our proposed online reputation monitoring system using opensource analytics tools involved three main processes which are data extraction, execution, and analysis as shown in Figure 1. Firstly, data extraction is the process of obtaining the social media contents in this case Facebook using an open source business analytics platform Microsoft Power BI. The second main process is the execution of sentiment analysis using selected text analytics API which is connected to a data analytics software platform, RapidMiner. Finally, the generated results can be exported into Microsoft Excel for analysis using various formulas. Visualization using Excel is also useful for better understanding and to support business decisions especially relation to online reputation.

4. Findings

In this section we present and discuss the findings and connect them with entire phases of the research.

4.1. Foundation phase outcome

Based on the conducted interview, we discovered that the organization does not have any preventive and curative strategy to use social media analytics for dealing with such online reputation attacks. However, they do invest in improving their organization's visibility on social media, e.g., through the promotion of achievements of the university and organized programs.

The key person quoted "We have not implemented any social media analytics so far for any purposes including Online Reputation Management like you mentioned, but we agree on the importance of online reputation protection…but we think it is not urgent at the moment and I believe it is a complex and costly thing". The quotation shows that the officer's perception towards the connection of ORM and the role of social media analytics in the job is not convincing. They perceived that social media analytics is a complex process, and is costly for acquiring the tools and services, and hiring staff e.g. data analysts. However, the key person expressed an interest to know what people are saying about his organization, either from the inside or outside. The interviewee quoted "But, we are interested to know how students respond to our posts on social media, so we always check their comments … negative comments are extremely low on our page, but we don't know if there are many on other pages". The lack of negative comments at an organization's social media account does not mean our online reputation is 100 percent positive. Many factors can contribute to the polarity such as internal social communication culture. In a university context, the students may be worried to comment negatively, but this cannot be generalized to all universities, as some are more open for criticism. Furthermore, online reputation as explained in [5] is defined as how people in the world are looking at us, not only from inside the organization. The issue on the banned universities is one example of how an organization's online reputation can be at risk. Thus, ORM with social media analytics provide preventive measure before a certain issue raised and also monitor an ongoing issue in news report to preserve reputation.

4.2. Experiment phase outcome

There are two tests in this phase which is accuracy test and sentiment analysis test. First, we want to observe and evaluate the accuracy of the selected open source text analytics APIs on product posts from Facebook. This is to know the reliability of the opensource API to be used in reputation monitoring. Second, we attempt to observe the sentiment polarity on news posted on Facebook that is predicted by the text analytics API. Figure 1 above shows the entire process involved in the proposed system. As we have measured the accuracy of the M API in experiment one, we believe that the API is reliable for experiment two. In the following, we described the results of the experiment according to the processes involved in the experiment in the next section.

4.3. Data extraction

In this study, we have extracted Facebook comments from two official Facebook pages. We have two different tests for different datasets extracted from Facebook. Dataset one consists of 1,122 lines of comments from three posts of a Facebook page of an international car company operating in Malaysia between May 1st to July 14th 2017. Meanwhile, dataset two consists of 89 comments from October 2nd to October 4th, 2017, from a local English online newspaper on the report of the banned universities.

The Facebook comments were extracted using Microsoft Power BI. We need to connect a specific Facebook account name to extract comments from it. There was a simple query in DAX (Data Analysis Expression) function language needed to be edited. Meanwhile, the speed of the extraction process depends on computer performance and the Internet connection. The extracted data were placed in an Excel worksheet for easy customization. For example, we can filter the extracted data based on our needs such as comments that contain certain words or comments posted at a certain date and time. The selected data were downloaded in CSV (comma separated values) format then exported in Microsoft Excel format (.xlsx) to be processed by RapidMiner Studio for sentiment polarity prediction.

Some comments needed to be removed for example those Malay, Chinese and India Language as well as comments that only consist of tagged Facebook ID and those in symbols or emoticons only. The API and RapidMiner generally support English and few other languages such as Spanish and German. We hired an advanced English user to annotate all 1,122 extracted comments with positive, 'negative', 'neutral' and 'other' for each comment. Other means that the comment needed to be removed. For dataset one, only 571 comments accepted, and 551 comments were removed. However, all 89 comments for dataset two were accepted.

4.4. Execution

In experiment one on the evaluation of the accuracy of selected API, we discovered that the results were not convincing but acceptable. The API predicted 305 comments or 0.60 of the total correct or similar with human annotated polarity. The matching process was done using Microsoft Excel's "EXACT" formula for text: '=EXACT(text1, text2)' where text1 = annotated text and text2=predicted text. If the text matched, e.g., text1 = negative, text2=negative, then it will print "TRUE". The count of TRUE and FALSE comments was done using COUNTIF formula: COUNTIF (range, criteria).

We observed that, polarity prediction of Facebook comments by the API are acceptable. This is because both manual and computerized prediction show majority of negative polarity. Manual annotation, records 286 and API predicted 214 of negative comments. The second highest polarity type by manual and API prediction is neutral, which is 199 and 207 comments. Based on the extracted comments, we suspected that the high amount of neutral comments predicted was due to the mixture of languages, short form of words, as well as sarcastic words. Meanwhile, the very low positive comments mean that the car company should do something to reduce the negative comments which can risk their online reputation. We can conclude that the customers were strongly dissatisfied with products from the car company.

Table 1. Results of Accuracy Test (Car Company) and Sentiment Analysis Test (Banned Universities Reports) by M Text Analytics API

		Accuracy Rate
Experiment 1	True	0.60 (331 comments)
	False	0.40 (220 comments)
		Sentiment Prediction
Experiment 2	Positive	0.07 (6 comments)
	Negative	0.12 (11 comments)
	Neutral	0.81 (72 comments)

In experiment two of banned universities news dataset, the API predicts a high number of neutral comments, as shown in Table 1. However, based on the comments from the news, we discovered another reason for the high amount of neutral comments is due to the tagging of other Facebook IDs. There is no comment written, but readers wanted to alert their contacts about the news. Nevertheless, negative comments are only slightly higher than positive comments which is to us not very impactful in our opinion.

The API also provides polarity confidence between 0 to1.0, where a number close or equal to 1.0 means a higher polarity, e.g., very negative comment. We filtered the data in Excel Online to select only the comments which have a polarity confidence of more than 7.0. One of three comments classified as 1.0 by API 1 sounds harsh and emotional as *"Shame Shame Where U put your s******d face Now?"*. We also filtered the lowest polarity confidence of negative classified comment, which has score of 0.475, and was quoted as *"*** is a joke university"*. The comment referred to one of the affected universities in a short form word. The comment looks cynical or sarcastic, and is not literally negative, but its meaning is negative. However, we supposed that collecting and combining more online news about certain issue with embedded Facebook comments can bring better results. Probably, the use web crawler is good option.

5. Implication for design

In the following, we conclude several advantages and limitations of the designed open source online reputation system.

5.1. Support in business decision making

Although the polarity prediction is not 100 percent accurate, it is acceptable for business decisions such as strategy development, marketing research and action plans. The prediction can be used to overview subjective opinions from customers toward certain issues, people, products, and services. This provides evidence-based reputation for organizations to allow them to manage it [3]. For instance, in of sentiment analysis on the car company above, proactive actions can be taken to reduce negative reputation. This is in line with [13], who mention that online reputation can improve competitiveness in the automotive industry. In case the organization participated in context study above, this approach can help it to analyse people's reactions towards the banned universities issue for appropriate action planning and online reputation recovery. Furthermore, they can go directly to negative comments or rumours which show a high polarity confidence and respond directly to the comments for rebuttal or denial.

5.2. Flexible monitoring strategies

We observed that the approach can be used for periodical monitoring, e.g., daily, weekly, or monthly; event-based monitoring, e.g., rumours or sudden issues; and entity monitoring, e.g. products, services and employees [11]. Periodical monitoring is probably a good preventive action to always keep pace with online reputation status. In case of the interviewed organization, event-based monitoring can be carried out to keep track of people's reactions towards the news.

5.3. Affordable for novice

In general, this approach can be applied to any organization or individual, e.g., celebrities or politicians. However, we suppose that large organizations prefer to outsource ORM services to analytics companies or acquire a sophisticated solution. The experiments above do not involve any cost, as all the tools used are open source with certain limits. We suggest that this approach is probably appropriate for small to medium sized profit-based or non-profit-based organizations or even individuals who are novice to ORM, but are interested to implement it [6]. It is also useful for academic researchers who are new to social media analytics and ORM.

6. Limitation

Despite the advantages of the approach, we discovered some limitations which are worthy to be shared. Among important limitations is the lack of Malay Language support by either the data science software platform, RapidMiner, or the text analytics API. This is because contents in the local language has a greater value to be analysed.

7. Conclusion and future work

This research attempts to propose an affordable and reliable ORM system using open source tools for any type and size of organization. The task can be a new role for public relation unit and it can be performed by a staff with IT background. The research also evaluates the performance of the system in term of prediction accuracy. We discovered that the accuracy of predictions by M text analytics API is acceptable (over 50 percent accurate). However, different text analytics API may produce different accuracy rate. This shows that the system is reliable to support business decisions and offers flexible monitoring purposes as preventive action. Monitoring strategies can be product or brand based or issue based. In future work, we want to extend the research on the evaluation of topic and entity extraction from user-generated contents (UGCs) using open source tools. We want to know what topics are involved in the conversation based on the extracted social media data.

References

[1] Jay B (Update) Oman Bans Four Malaysian Universities Over Alleged ‘Academic, Administrative Abuse, Kuala Lumpur, 02-Oct-2017
[2] Bernama “USIM Says Administrative Issues to Blame for Oman Ban,” Seremban, pp. 6–11, 06-Oct-2017
[3] Aula P and Heinonen J 2016 Evidence-based reputation management Reputable Firm. Manag. Prof., 1, pp 163–199
[4] E. Commentary 2017 “Management Mitigating Risks through Online Reputation,” pp. 1–7
[5] Conner C 2014 “Top Online Reputation Management Tips for Brand Marketers,” Forbes
[6] Hartzer B 2017 “How Much Does Online Reputation Management Cost?” Billhartzer.com, [Online]. Available: https://www.billhartzer.com [Accessed: 29-Oct-2017]
[7] Rouse M 2017 “What is social media analytics?” searchbusinessanalytics.techtarget.com [Online]. Available: https://www.cio.com/article/3185725/security/9-biggest-information-security-threats-through-2019.html. [Accessed: 09-Sep-2017]
[8] Khan G 2015 Seven Layers of Social Media Analytics: Mining Business Insights from Social Media Text, Actions, Networks, Hyperlinks, Apps, Search Engine, and Location Data 1st ed. (CreateSpace Independent Publishing Platform) chapter 1 p 1
[9] Segerberg A and Bennett L 2011 Social media and the organization of collective action : using twitter to explore the ecologies of two climate change protests social media and the organization of collective action Commun. Rev. 14 3 pp 197–215
[10] Capatina A and Schin 2017 Perceptions on Online Reputation Management in Tourism Industry (The Case of a Romanian Hotel) J. Manag. Res. 9 2 p 71
[11] Guo S, Wang M and Leskovec J 2011 The Role of Social Networks in Online Shopping: Info. Passing, Price of Trust, and Consumer Choice Proc. ACM Conf. E-Commer pp 157–166
[12] Dorčák P and František P 2015 Analysis of the possibilities of improving an online reputation of public institutions J. Soc. Sci. 4 1 pp 275–282
[13] Dorčák P, Peter M, and Pollák F 2015 Multifactor analysis of online reputation as a tool for enhancing competitiveness of subjects from automotive industry Ekon. 65 2 pp 173–186
[14] Hussein D 2016 A survey on sentiment analysis challenges J. King Saud Univ. - Eng. Sci.
[15] Mejova Y 2012 Sentiment analysis within and across social media streams p. 190
[16] Kharde V and Sonawane 2016 Sentiment analysis of twitter data : A survey of techniques Int. J. Comput. Appl. 139 11 pp. 5–15
[17] Laeeq F and Tabrez N 2017 Sentimental Classification of Social Media using Data Mining Int. J. of Adv. Res. in Comp. Sci. (Udaipur) 8 5 pp 546-549
[18] Liao W and Chhoeut R 2017 Using Social Networking APIs to Collect Real-time Data for Analytics Social Proc. West Virginia Acad. Sci. vol. 89, no. 1
[19] Sathish R, Ambika M, Prabhu R and Krishnakumar 2017 R Real time data collection and

analytics of social media sites using netlytic Middle-East J. Sci. Res. 25 1 pp 174–180

[20] Lomborg S and Bechmann A 2014 Using APIs for Data Collection on Social Media Using APIs for Data Collection on Social Media Int. J. Inf. Soc. 30 4 256–265

[21] Dehghani M, Johnson K, Garten J, Boghrati R, Hoover J, Balasubramanian V, Singh A and Shankar Y 2016 TACIT : An open-source text analysis, crawling, and interpretation tool Beh. Res. Met. 49 2 pp 538-547

[22] Pandita R, Jetley R, Sudarsan S, Menzies T and L. Williams 2017 TMAP: Discovering Relevant API Methods Through Text Mining Of API Documentation Special Issue SCAM

A Systematic Literature Mapping of Risk Analysis of Big Data in Cloud Computing Environment

Hazirah Bee Yusof Ali, Lili Marziana Abdullah, Mira Kartiwi, Azlin Nordin, Norsaremah Salleh, Normi Sham Awang Abu Bakar

Kulliyah of Information and Communication Technology, International Islamic University Malaysia,
hazirahbee@gmail.com, lmarziana@iium.edu.my, mira@iium.edu.my, azlinnordin@iium.edu.my, saremah@iium.edu.my, normi@iium.edu.my

Abstract This paper investigates previous literature that focusses on the three elements: risk assessment, big data and cloud. We use a systematic literature mapping method to search for journals and proceedings. The systematic literature mapping process is utilized to get a properly screened and focused literature. With the help of inclusion and exclusion criteria, the search of literature is further narrowed. Classification helps us in grouping the literature into categories. At the end of the mapping, gaps can be seen. The gap is where our focus should be in analysing risk of big data in cloud computing environment. Thus, a framework of how to assess the risk of security, privacy and trust associated with big data and cloud computing environment is highly needed.

1. Introduction

The emergence of recent technologies changes the way we live. We spend more time chatting and discussing with each other virtually. Accumulation of data from continuous chatting and discussions makes the data bigger. Thus, the best location would always be the cloud. The use of cloud as a storage not only reduces the cost of operation but also increases the performance of companies. Other than using cloud as a storage, cloud computing provides companies with a shared infrastructure, shared platform and shared software. The shared environment promotes portability whereby regardless of storage location, companies can still operate efficiently as before. Big data in cloud computing reduces cost of IT facilities and lowers the capital investment of companies [1]. Companies today become more effective and efficient in daily operation compared to when cloud computing is not in existence.

However, the increasing rate of cloud usage by companies forces us to think about the negative side of cloud computing [2]. If we use cloud in our work, we should also be aware of the risks and consequences. We should be able to protect ourselves from that risk [2]. This paper presents our investigation of the literature using the systematic mapping method to understand what research has been done on risk analysis of big data in the cloud environment. Systematic Mapping method enables us to search the literature of our research focus systematically and thoroughly. Section 2 presents the mapping method and discusses the steps taken in the process. Section 3 answers the mapping question. Section 4 presents the classification and Section 5 concludes the research by highlighting the gap.

IOP Conf. Series: Journal of Physics: Conf. Series **1018** (2018) 012017 doi:10.1088/1742-6596/1018/1/012017

2. Method

There are two types of literature review method which are narrative and systematic. Narrative or traditional method is descriptive and has few search techniques; thus, may lead to biased results [3]. Systematic is structured around focused questions and has less tendency to create bias in its results [3]. The systematic literature mapping presented in this paper was done prior to our systematic literature review of relevant papers. Systematic mapping helps us to search for meaningful literature so that we can produce summarized view of the research area. The mapping process starts with the formulation of mapping questions. Using accurate search strings in academic search databases, only relevant literature is captured. Through the screening of literature and keywording of abstract, the literature is mapped into meaningful classifications that relate to the focus of study. The mapping shows the gaps in the research. We adopted systematic mapping process done by [4] and the steps taken are listed below.

- Define the Research Questions
- Conduct the Search
- Screening of Papers
- Keywording using Abstracts
- Data Extraction and Mapping Process

2.1 Design the Research Questions

The mapping process starts with the formulation of mapping questions to know about previous literature on risk analysis of big data in cloud computing environment. As there may have been studies done on the area, we would like to have to an overview of the research that has or has not been done. To achieve our intention, we formulated the following research questions:

MQ1: What are the studies that examined risks analysis of big data and cloud computing?
MQ2: Which publications are mostly used for research production in the area?
MQ3: What are the most used research approaches in the area?

2.2 Conduct the Search

We selected several academic databases that are Computer Science, and (ICT) Information and Communication Technology indexed. The selected databases are ACM, IEEE Explore, Science Direct, and Springer Link, which are subscribed by the university; thus, enabling us to retrieve the literature with ease. We conduct the search using keywords that include synonyms and common words so that we can answer the research questions. The search string chosen for the search:

> ("risk assessment" OR "risk analysis" OR "risk management" OR "risk treatment" OR "risk mitigation") AND "big data" AND cloud

The use of " " is to ensure that we have the literature containing both words. The use of OR is to ensure that we get any of the five terms. The use of AND is to ensure that we get all three keywords. "risk assessment", "risk analysis", "risk management", "risk treatment" and "risk mitigation" are common terms used in the risk research area and are sometimes used interchangeably. Using the selected search string at several databases returned the following number of literature as shown in Table 1. We tested the search string on IEEE Explore and after checking the validity of the literature produced, we decided to use the same search technique for other search databases. The search results with high numbers of literature demonstrate the thorough coverage of the search strings.

Table 1: Online Database Search results

Online Database	Search results
IEEE Explore	698
ACM	633
Science Direct	471
Springer Link	46
	1848

2.3 Screening of Papers
The screening of papers involved two phases so that the papers selected are more focused to our research area. The selection was based on the inclusion and exclusion criteria described below.

2.31 Phase 1
We filter the papers by reading the Topic and Abstract. We need to identify whether the papers are appropriate for the research. The Topic and Abstract should be referring to risk, big data and cloud. Table 2 shows that the number of papers have been reduced in this phase as several were removed.

2.32 Phase 2
In Phase 2, we screened the papers further by reading the Introduction, Method and Conclusion. These three sections highlighted their suitability for the research. At the end of Phase 2, we managed to reduce the number of papers to 56 papers. The literature resulted from Phase 2 focuses on our keywords not just in Title and Abstract but also in Introduction, Method and Conclusion. Therefore, we can claim that we have the most appropriate literature for our research.

2.33 Inclusion Criteria
All papers that relate to risk, big data and cloud were included in the research. The papers must contain empirical data to support the findings [4].

2.34 Exclusion Criteria
All papers that are not related to risk, big data and cloud were excluded. Papers that provide opinions and have no empirical data were also excluded [4].

Table 2: Online Database Search Results for Phase1 and Phase2

Online Databases	Search results	Phase 1	Phase 2
IEEE Explore	698	181	50
ACM	633	72	4
ScienceDirect	471	29	1
Springer link	46	6	1
TOTAL	1848	288	56

2.4 Keywording of Abstract
Keywording of abstract is the activity of looking for relevant keywords manually from the abstract [4]. This process reduces our time in looking for the correct classifications [5]. The themes arising from the keywords in the abstract allows us to segregate the literature into five classifications which are Risk Identification, Risk Analysis, Risk Mitigation, Risk Monitoring and Control and Risk Solution.

2.5 Data Abstraction and Mapping of Studies
At this stage, we then mapped each literature into the classification scheme shown below that we built through the keywording of abstracts. The scheme is meant to help us answer the mapping questions.

- Risk Identification
- Risk Assessment
- Risk Mitigation
- Risk Monitoring and Control
- Risk Solution

3. Answer to the Mapping Question

MQ1: What are the studies that examined risks analysis of big data and cloud computing?
There are 56 papers that examined risks analysis of big data and cloud computing. We sorted the papers into five classifications: Risk Identification, Risk Assessment, Risk Mitigation, Risk Monitoring and Control and Risk Solution. 12 papers discussed Risk Identification, six papers discussed Risk Assessment, 12 papers discussed Risk Mitigation, 11 papers discussed Risk Monitoring and Control, and finally 10 papers discussed Risk Solution. These papers are highlighted in Section 4 and included in the reference section.

MQ2:Which publications are mostly used for research production in the area?
Out of 56 papers, 17 papers are journal papers while the rest are from proceedings of conferences. The journals are published by IEEE ranging from IEEE Software, IEEE Security and Privacy and IEEE. The proceedings are mostly from IEEE conferences. Few proceedings originate from ACM.

MQ3:What are the most used research approaches in the area?
We grouped the approaches used into four research approaches; theoretical, experimental, simulation and modeling [3]. These are research approaches meant for computer science research. 28 papers used theoretical approach, nine papers used experimental approach, seven papers used simulation approach and 12 papers used modeling approach. Theoretical approach covers 49% of the literature.

4. Classifications

In risk management, risk is usually classified in this order: risk identification, risk analysis, risk response plan, risk monitoring and management [6]. For this research, we classified the literature in such a way that it helps us to answer the research questions. We added another class in the classification: risk identification, risk assessment, risk mitigation, risk monitoring and control and finally risk solution.

4.1 Risk Identification

Risk identification are activities taken to identify, to determine and to detect risk for big data in cloud. Literature works referring to the identification, determination and detection were grouped in risk identification. 12 papers discussed risk identification and most of them focused on risk detection [7][8][9][10][11][12][13][14][15][16][14][17].

4.2. Risk Assessment

Risk assessment are activities taken to estimate and prioritize the risks of using big data in the cloud environment[18]. In risk assessment, risks are analysed and investigated. Risk assessment can be done after risk has been identified. Six papers focused on risk assessment [19][18][20][21][22][23].

4.3. Risk Mitigation

Risk mitigation are steps or activities taken to reduce the risks [24]. Risk mitigation is conducted after risk has been identified and assessed. At this stage, we already know about the risk. We must find ways either to avoid the risk or to reduce the risk. 12 papers discussed risk mitigation [25][26][27][28][29][30][2][31][32][33][34][24].

4.4. Risk Monitoring and Control

Risk Monitoring and Control are activities to monitor identified risks and to ensure that the risks can be controlled. 15 papers fell under this classification and most of them focused on risk management framework or model [35] [36][37][38][39][40][41][42][43][44][45][46][47][48][49].

4.5. Risk Solution

Risk Solution are activities taken to solve risk issues [29]. In this classification, most of the literature focused on solutions such as algorithm and prototypes. 11 papers discussed on risk solution [50][51][52][53][54][55][56][57][58][59][60].

5. Gap Identification

Table 3: Classifications vs Research Approaches

	Theoretical	Experimental	Simulation	Modeling	Total
Risk Identification	3	2	2	5	12
Risk Assessment	3	2	0	1	6
Risk Mitigation	6	2	3	1	12
Risk Monitoring and Control	11	2	1	1	15
Risk Solution	5	1	1	4	11
Total	28	9	7	12	56

Table 3 above shows that Risk Monitoring and Control has the highest number of literature, followed by Risk Identification, Risk Mitigation, Risk Solution and finally Risk Assessment. Less number of literature for risk assessment indicates that not many researchers are researching on risk assessment. One paper focused on risk for mobile ad hoc network [21], one paper focused on risk assessment for ITL at NIST [18] and another one paper focused on risk assessment for cyber physical power system [22]. Systematic literature review was done for security and privacy for big data [61]. One paper discussed the qualitative and quantitative research for risk assessment in cloud [23] and one paper introduced cloud model into information security risk assessment [19].

6. Conclusion

There is a gap in risk assessment as clearly stated in Table 3. Only six papers talked about risk assessment compared to more than ten papers that talked about risk identification, risk mitigation, risk monitoring and control and risk solution. Therefore, we can conclude that there is a need for research in risk assessment for big data in cloud computing environment.

Systematic mapping activities do help us see the gap. The process of implementing them makes us more confident since they are fully supported by the literature. Accumulated literature is geared towards answering the research questions because the search of literature follows a systematic method and is well documented.

Acknowledgements

This research has been funded by Fundamental Research Grant Scheme (FRGS16-022-0521) supported by Ministry of Higher Education.

References

[1] I. A. T. Hashem, I. Yaqoob, N. B. Anuar, S. Mokhtar, A. Gani, and S. Ullah Khan, "The rise of 'big data' on cloud computing: Review and open research issues," *Inf. Syst.*, 2015.

[2] Y. S. Rathore, "Measuring Cloud Security from risks," pp. 214–220, 2016.

[3] J. D. Baker, "The Purpose, Process, and Methods of Writing a Literature Review," *AORN J.*, vol. 103, no. 3, pp. 265–269, 2016.

[4] S. Zein, N. Salleh, and J. Grundy, "A systematic mapping study of mobile application testing techniques," *J. Syst. Softw.*, vol. 117, no. July, pp. 334–356, 2016.

[5] S. Zein, N. Salleh, and J. Grundy, "A systematic mapping study of mobile application testing techniques," *J. Syst. Softw.*, vol. 117, pp. 334–356, 2016.

[6] V. G. Semin and E. G. Shmakova, "The Information Security Risk Management," pp. 106–109, 2017.

[7] C. P. Udeagwu, S. Sotiriadis, E. Asimakopoulou, N. Bessis, and M. Trovati, "Analysis of Techniques for Visualizing Security Risks and Threats," *Proc. - 2015 10th Int. Conf. P2P, Parallel, Grid, Cloud Internet Comput. 3PGCIC 2015*, pp. 584–590, 2015.

[8] G. Collard, E. Disson, G. Talens, and S. Ducroquet, "Proposition of a method to aid Security Classification in Cybersecurity context," *2016 14th Annu. Conf. Privacy, Secur. Trust. PST 2016*, pp. 88–95, 2016.

[9] O. Johny, S. Sotiriadis, E. Asimakopoulou, and N. Bessis, "Towards a social graph approach for modeling risks in big data and internet of things (IoT)," *6th Int. Conf. Intell. Netw. Collab. Syst. IEEE INCoS 2014*, vol. 26, pp. 439–444, 2014.

[10] H. Kikuchi, T. Yamaguchi, K. Hamada, Y. Yamaoka, H. Oguri, and J. Sakuma, "Ice and fire: Quantifying the risk of re-identification and utility in data anonymization," *Proc. - Int. Conf. Adv. Inf. Netw. Appl. AINA*, vol. 2016–May, pp. 1035–1042, 2016.

[11] Z. . Salek and F. M. . Madani, "Multi-level Intrusion detection system in cloud environment based on trust level," *2016 6th Int. Conf. Comput. Knowl. Eng. ICCKE 2016*, no. Iccke, pp. 94–99, 2016.

[12] K. Vassiljeva, A. Tepljakov, E. Petlenkov, and E. Netšajev, "Computational Intelligence Approach for Estimation of Vehicle Insurance Risk Level," pp. 4073–4078, 2017.

[13] Y.-K. Kim and C.-S. Jeong, "Risk Prediction System Based on Risk Prediction Model with Complex Event Processing: Risk Prediction in Real Time on Complex Event Processing Engine," *2014 IEEE Fourth Int. Conf. Big Data Cloud Comput.*, pp. 711–715, 2014.

[14] K. D. B. H. Subasinghe and S. R. Kodithuwakku, "A big data analytic identity management expert system for social media networks," *2015 IEEE Int. WIE Conf. Electr. Comput. Eng. WIECON-ECE 2015*, pp. 126–129, 2016.

[15] M. He, H. Ji, Q. Wang, and C. Ren, "Big data fueled process management of supply risk: Sending, prediction, evaluation and mitigation," *Proc. 2014 Winter Simul. Conf.*, pp. 1005–1013, 2013.

[16] Z. Lianbing, "Study on Applying the Neural Network in Computer Network Security Assessment," *2016 Eighth Int. Conf. Meas. Technol. Mechatronics Autom.*, no. 2, pp. 639–642, 2016.

[17] L. R. Techio and M. Misaghi, "EMSCLOUD - An evaluative model of cloud services cloud service management," *5th Int. Conf. Innov. Comput. Technol. INTECH 2015*, no. Intech, pp. 100–105, 2015.

[18] Lulu Liang, Wang Ren, Jing Song, Huaming Hu, Qiang He, and Shuo Fang, "The state of the art of risk assessment and management for information systems," *2013 9th Int. Conf. Inf. Assur. Secur.*, pp. 66–71, 2013.

[19] D. Zong-you, Z. Wen-long, S. Yan-an, and W. Hai-tao, "The application of cloud matter — Element in information security risk assessment," *2017 3rd Int. Conf. Inf. Manag.*, pp. 218–222, 2017.

[20] B. Nelson and T. Olovsson, "Security and privacy for big data: A systematic literature review," *2016 IEEE Int. Conf. Big Data (Big Data)*, pp. 3693–3702, 2016.

[21] S. Tanimoto, S. Shiraki, M. Iwashita, H. Sato, T. Kobayashi, and A. Kanai, "Risk Assessment Based on User's Viewpoint for Mobile Ad Hoc Network," *NBiS 2016 - 19th Int. Conf. Network-Based Inf. Syst.*, pp. 280–285, 2016.

[22] S. Deng *et al.*, "Unattended Remote Attestation Delegation for Grid Computing," *IEEE Trans. Intell. Transp. Syst.*, vol. 8, no. 3, pp. 229–238, 2012.

[23] V. Bellandi, S. Cimato, E. Damiani, G. Gianini, and A. Zilli, "Toward Economic-Aware Risk Assessment on the Cloud," *IEEE Secur. Priv.*, vol. 13, no. 6, pp. 30–37, Nov. 2015.

[24] H. A. Kholidy, "Cloud Computing Systems," 2015.

[25] E. Damiani, "Toward big data risk analysis," *Big Data (Big Data), 2015 IEEE Int. Conf.*, pp. 1905–1909, 2015.

[26] H. M. Chen, R. Kazman, and S. Haziyev, "Strategic prototyping for developing big data systems," *IEEE Softw.*, vol. 33, no. 2, pp. 36–43, 2016.

[27] A. Khosravani, "A case study analysis of risk, trust and control in cloud computing," ... *Conf. (SAI), 2013*, pp. 879–887, 2013.

[28] M. Felici and S. Pearson, "Accountability, Risk, and Trust in Cloud Services: Towards an Accountability-Based Approach to Risk and Trust Governance," *2014 IEEE World Congr. Serv.*, pp. 105–112, 2014.

[29] W. Wongthai, F. Rocha, and A. Van Moorsel, "Logging solutions to mitigate risks associated with threats in infrastructure as a service cloud," *Proc. - 2013 Int. Conf. Cloud Comput. Big Data, CLOUDCOM-ASIA 2013*, pp. 163–170, 2013.

[30] A. A. Almutairi and A. Ghafoor, "Risk-Aware Virtual Resource Management for Multitenant Cloud Datacenters," *IEEE Cloud Comput.*, vol. 1, no. 3, pp. 34–44, Sep. 2014.

[31] N. Miloslavskaya, "Security operations centers for information security incident management," *Proc. - 2016 IEEE 4th Int. Conf. Futur. Internet Things Cloud, FiCloud 2016*, pp. 131–138, 2016.

[32] S. Supriya and S. Padaki, "Data Security and Privacy Challenges in Adopting Solutions for IOT," *Proc. - 2016 IEEE Int. Conf. Internet Things; IEEE Green Comput. Commun. IEEE Cyber, Phys. Soc. Comput. IEEE Smart Data, iThings-GreenCom-CPSCom-Smart Data 2016*, pp. 410–415, 2017.

[33] K. Gai, M. Qiu, and S. A. Elnagdy, "A Novel Secure Big Data Cyber Incident Analytics Framework for Cloud-Based Cybersecurity Insurance," *Proc. - 2nd IEEE Int. Conf. Big Data Secur. Cloud, IEEE BigDataSecurity 2016, 2nd IEEE Int. Conf. High Perform. Smart Comput.*

IEEE HPSC 2016 IEEE Int. Conf. Intell. Data S, pp. 171–176, 2016.

[34] A. Mtibaa, K. A. Harras, and H. Alnuweiri, "Malicious attacks in Mobile Device Clouds: A data driven risk assessment," *Proc. - Int. Conf. Comput. Commun. Networks, ICCCN*, 2014.

[35] D. Mellado and E. Fernández-medina, "A Systematic Review of Information Security Governance Frameworks in the Cloud Computing Environment," *2016 IEEE 18th Int. Conf. High Perform. Comput. Commun.*, vol. 18, no. 6, pp. 798–815, 2012.

[36] V. Agrawal, "A Framework for the Information Classification in ISO 27005 Standard," *Proc. - 4th IEEE Int. Conf. Cyber Secur. Cloud Comput. CSCloud 2017 3rd IEEE Int. Conf. Scalable Smart Cloud, SSC 2017*, pp. 264–269, 2017.

[37] S. A. Elnagdy, M. Qiu, and K. Gai, "Understanding Taxonomy of Cyber Risks for Cybersecurity Insurance of Financial Industry in Cloud Computing," *Proc. - 3rd IEEE Int. Conf. Cyber Secur. Cloud Comput. CSCloud 2016 2nd IEEE Int. Conf. Scalable Smart Cloud, SSC 2016*, pp. 295–300, 2016.

[38] M. Sun, "Big Data : Risks and Regulatory Strategies," pp. 358–362, 2015.

[39] X. Z. Wang, X. L. Bi, and Z. Q. Ge, "Deep Data Fusion Model for Risk Perception and Coordinated Control of Smart Grid," *2015 Int. Conf. Estim. Detect. Inf. Fusion (ICEDIF 2015)*, no. ICEDIF, pp. 0–3, 2015.

[40] T.-M. Choi, H. K. Chan, and X. Yue, "Recent Development in Big Data Analytics for Business Operations and Risk Management," *IEEE Trans. Cybern.*, vol. 47, no. 1, pp. 81–92, 2017.

[41] H. Jean-Baptiste, M. Qiu, K. Gai, and L. Tao, "Model Risk Management Systems-Back-End, Middleware, Front-End and Analytics," *Proc. - 2nd IEEE Int. Conf. Cyber Secur. Cloud Comput. CSCloud 2015 - IEEE Int. Symp. Smart Cloud, IEEE SSC 2015*, pp. 312–316, 2016.

[42] W. Abbass, A. Baina, and M. Bellafkih, "Improvement of information system security risk management," *2016 4th IEEE Int. Colloq. Inf. Sci. Technol.*, pp. 182–187, 2016.

[43] A. Chen, H. Xing, K. She, and G. Duan, "A Dynamic Risk-Based Access Control Model for Cloud Computing," *2016 IEEE Int. Conf. Big Data Cloud Comput. (BDCloud), Soc. Comput. Netw. (SocialCom), Sustain. Comput. Commun.*, no. 2014, pp. 579–584, 2016.

[44] S. J. Yang and I. C. Cheng, "Design Issues of Trustworthy Cloud Platform Based on IP Monitoring and File Risk," *Proc. - 2015 IEEE 5th Int. Conf. Big Data Cloud Comput. BDCloud 2015*, pp. 110–117, 2015.

[45] M. S. K. Awan, P. Burnap, O. Rana, and A. Javed, "Continuous monitoring and assessment of cybersecurity risks in large computing infrastructures," *Proc. - 2015 IEEE 17th Int. Conf. High Perform. Comput. Commun. 2015 IEEE 7th Int. Symp. Cybersp. Saf. Secur. 2015 IEEE 12th Int. Conf. Embed. Softw. Syst. H*, pp. 1442–1447, 2015.

[46] K. Hogan *et al.*, "Secure Multiparty Computation for Cooperative Cyber Risk Assessment," *Proc. - 2016 IEEE Cybersecurity Dev. SecDev 2016*, pp. 75–76, 2017.

[47] D. Zhe, W. Qinghong, S. Naizheng, and Z. Yuhan, "Study on Data Security Policy Based on Cloud Storage," *2017 IEEE 3rd Int. Conf. Big Data Secur. Cloud (BigDataSecurity), IEEE Int. Conf. High Perform. Smart Comput. IEEE Int. Conf. Intell. Data Secur.*, pp. 145–149, 2017.

[48] M. Paryasto, A. Alamsyah, and B. Rahardjo, "Big-data security management issues," *2014 2nd Int. Conf. Inf. Commun. Technol.*, pp. 59–63, 2014.

[49] M. Coelho and A. Rau-Chaplin, "EXsight: An analytical framework for quantifying financial loss in the aftermath of catastrophic events," *Proc. - Int. Work. Database Expert Syst. Appl.*

DEXA, pp. 165–170, 2014.

[50] M. Abdlhamed, K. Kifayat, Q. Shi, and W. Hurst, "A System for Intrusion Prediction in Cloud Computing," *Proc. Int. Conf. Internet Things Cloud Comput.*, p. 35:1--35:9, 2016.

[51] Q. Liu, E. C.-H. Ngai, X. Hu, Z. Sheng, V. C. M. Leung, and J. Yin, "SH-CRAN: Hierarchical Framework to Support Mobile Big Data Computing in a Secure Manner," *Proc. 2015 Work. Mob. Big Data*, pp. 19–24, 2015.

[52] J. Whitworth and S. Suthaharan, "Security problems and challenges in a machine learning-based hybrid big data processing network systems," *ACM SIGMETRICS Perform. Eval. Rev.*, vol. 41, no. 4, pp. 82–85, 2014.

[53] S. Hassan and G. Pernul, "Efficiently Managing the Security and Costs of Big Data Storage Using Visual Analytics," *Proc. 16th Int. Conf. Inf. Integr. Web-based Appl. & Serv.*, no. Dm, pp. 180–184, 2014.

[54] R. Alsufyani and V. Chang, "Risk Analysis of Business Intelligence in Cloud Computing," *2015 IEEE 7th Int. Conf. Cloud Comput. Technol. Sci.*, pp. 558–563, 2015.

[55] G. Hatzivasilis, I. Papaefstathiou, and C. Manifavas, "Software security, privacy, and dependability: Metrics and measurement," *IEEE Softw.*, vol. 33, no. 4, pp. 46–54, 2016.

[56] K. Gai, M. Qiu, and S. A. Elnagdy, "Security-Aware Information Classifications Using Supervised Learning for Cloud-Based Cyber Risk Management in Financial Big Data," *Proc. - 2nd IEEE Int. Conf. Big Data Secur. Cloud, IEEE BigDataSecurity 2016, 2nd IEEE Int. Conf. High Perform. Smart Comput. IEEE HPSC 2016 IEEE Int. Conf. Intell. Data S*, pp. 197–202, 2016.

[57] H. Jean-Baptiste, M. Qiu, K. Gai, and L. Tao, "Meta Meta-Analytics for Risk Forecast Using Big Data Meta-Regression in Financial Industry," *Proc. - 2nd IEEE Int. Conf. Cyber Secur. Cloud Comput. CSCloud 2015 - IEEE Int. Symp. Smart Cloud, IEEE SSC 2015*, pp. 272–277, 2016.

[58] R. Trapero, J. Luna, and N. Suri, "Quantifiably Trusting the Cloud: Putting Metrics to Work," *IEEE Secur. Priv.*, vol. 14, no. 3, pp. 73–77, 2016.

[59] B. Varghese and A. Barker, "Are Clouds Ready to Accelerate Ad Hoc Financial Simulations?," *Proc. - 2014 Int. Symp. Big Data Comput. BDC 2014*, pp. 54–63, 2015.

[60] A. M. Algarni and Y. K. Malaiya, "A consolidated approach for estimation of data security breach costs," *Proc. 2016 Int. Conf. Inf. Manag. ICIM 2016*, pp. 26–39, 2016.

[61] S. Li and J. Gao, "Security and Privacy for Big Data," in *Big Data Concepts, Theories, and Applications*, Cham: Springer International Publishing, 2016, pp. 281–313.

IOP Conf. Series: Journal of Physics: Conf. Series **1018** (2018) 012018 doi:10.1088/1742-6596/1018/1/012018

Range wise busy checking 2-way imbalanced algorithm for cloudlet allocation in cloud environment

Mohammed Alanzy[1] , Rohaya Latip[1,2], Abdullah Muhammed[1]

[1]Computer Sciences and Information Technology, Universiti Putra Malaysia,
[2]Institut Penyelidikan Matematik (INSPEM), Universiti Putra Malaysia,
43400 Serdang, Malaysia

m.alanzy.cs@gmail.com,
rohayalt@upm.edu.my,
abdullah@upm.edu.my.

Abstract. Cloud computing considers as a new business paradigm and a popular platform over the last few years. Many organizations, agencies, and departments consider responsible tasks time and tasks needed to be accomplished as soon as possible. These agencies counter IT issues due to the massive arise of data, applications, and solution scopes. Currently, the main issue related with the cloud is the way of making the environment of the cloud computing more qualified, and this way needs a competent allocation strategy of the cloudlet, Thus, there are huge number of studies conducted with regards to this matter that sought to assign the cloudlets to VMs or resources by variety of strategies. In this paper we have proposed range wise busy checking 2-way imbalanced Algorithm in cloud computing. Compare to other methods, it decreases the completion time to finish tasks' execution, it is considered the fundamental part to enhance the system performance such as the makespan. This algorithm was simulated using Cloudsim to give more opportunity to the higher VM speed to accommodate more Cloudlets in its local queue without considering the threshold balance condition. The simulation result shows that the average makespan time is lesser compare to the previous cloudlet allocation strategy.

1. Introduction

Cloud computing is the latest trend in computer science, and it is expected to be the future of the 21st century technology. Cloud computing is mostly common for its deployment of shared resources proficiently. The distribution of the cloudlets to the fittest resource, also known as virtual machines, is one of the main requirements in the environment of cloud computing [1]. A typical cloud environment holds a module called Data Centre Broker which carries out the controlling of the entire Data Centre, including the allocation of the cloudlets to the virtual machines. Similar to most computing performances, enhancement and optimization of the allocation algorithm will always be an open option. To engineer an effective cloudlet, allocation algorithm is considered to be a challenging research area [2]. Numerous policies have been suggested, examined and compared on heterogeneous parallel computing environments. The introduction of a new mechanism was made; effective aggregated computing power (EACP) which improves and enhances the performance [3]. On the other hand, the adaptive weighted factoring (AWF) is responsible for scheduling parallel loops [4]. The active loop scheduling via strengthened learning is helpful and effective the most in time stepping scientific platforms with numerous steps [5]. The scheduling strategies for grid surroundings take advantage of a few methods that are alike, yet, some difference to the mechanism of cloudlet allocation strategies are witnessed [6].

Genetic algorithms are also used for scheduling [7], the opportunistic load balancing heuristic randomly picks one of the batch's cloudlets and distributes it to the next virtual machine predicted to be available, not taking into consideration the cloudlet's anticipated completing time on that virtual machine, causing a very deprived makespan [8][9]. The minimum execution time (MET) distributes every cloudlet picked randomly to the virtual machine that has the lowest execution time causing the inequality of load through the virtual machines [10]. Minimum completion time (MCT) heuristic, distributes every cloudlet to the virtual machine with the least execution time for the cloudlet [10]. It figuratively and practically merges the advantages of OLB and MET. The quality of service guided Min-Min allocates cloudlets that need more bandwidth. The QOS priority grouping scheduling presents the significance of deadlines [11]. The Grid-JQA scheduling solution takes advantage of a combination formula which chains the parameters together along with weighting factors, in order to evaluate QOS [12][13]. Alternative user oriented scheduling algorithm that takes advantage of using progressive reservation and resource assortment methods reduces the completion time of single cloudlets without bearing in mind the makespan [14]. The multiple resource scheduling (MRS) algorithm studies the abilities of the system and the resource necessities of cloudlets as common aspects [15]. Within cloud computing settings, the majority of the distribution regulations load some precise resources moderately more heavily, parting other resources either idle or with the minimum burden [16]. As a cause, load balancing is a critical concern in cloud computing which touches the performance of the cloud service supplier. The objective of this work is to develop and improve the current allocation strategies in this area by creating a new cloudlet allocation strategy based on (RB2B) algorithm [17] which mainly concentrates on decreasing the makespan, and simultaneously, it optimizes the virtual machine utilization to a notable amount by allocating the cloudlets to the virtual machines in the best way possible or most fairness way. The proposed algorithm is merged in the Data Centre Broker (DCB) module.

2. Related Works

Some resource allocation polices that consider the cornerstone for the proposed algorithm, described as follow:

2.1. *Min-Min*

Min-Min [18][19][20] at first, a matrix is occupied for all the free cloudlets. Two phases occur in Min-min, initially, the first phase, the set of minimum computing time for every single cloudlet within the matrix is calculated and tackled. As for the second phase, the cloudlet with the total least computation time is picked from the matrix and allocated to the analogous virtual machine. After that, the given cloudlet is taken out from the matrix and admissions of the matrix are improved consequently. The course of Min-min would be recurrent until the matrix runs out of cloudlets, thus, the cloudlets within the matrix are all mapped. It takes $0(mn^2)$ time for this algorithm, where m is the number of virtual machines and n is the number of cloudlets. The Min-min algorithm has a lack in the utilization of the uniform resource due to select smaller Cloudlets at the beginning which makes use of the virtual machine with higher MIPS. Thus, the scheduling is not performing when the number with larger Cloudlets is less than the number of the small Cloudlets.

2.2. *Max-Min*

This algorithm [18] is very much like Min-min. However, there is a difference that stands out within the second phase. In Max-min, it first picks the cloudlet with the highest computation time from the matrix and allocates it to the virtual machine that has the selected cloudlet producing minimum time to complete. Like the previous algorithm Min-min, this algorithm also takes $0(mn^2)$ time where m and n are numbers of virtual machines and number of cloudlets respectively. To solve the weakness of Min-min, Max-min algorithm is introduced that firstly executes the larger Cloudlet. However, the overall makespan may increase as well as the cloudlets with smaller length and its waiting time will increase.

2.3. *Resource Aware Scheduling Algorithm (RASA)*

The algorithm RASA [18]combinations of both Min-min and Max-min algorithms. If the virtual machines count is odd, the Min-min algorithm is applied, otherwise, Max-min is applied. The entire process may be distributed into a number of sequences, where two cloudlets are distributed to suitable virtual machines by any of the two policies, nonetheless. The rule dictates if the first cloudlet of current round is distributed to a virtual machine using the strategies of Min-min, the following cloudlet would be distributed by the max-min strategy. In the upcoming round, the allocation of the cloudlet initiates with an algorithm unlike the one from the previous round, to make it clearer, if the first round begins with max-min, the other initiates with Min-min and vice versa. The experimental results indicate that if the amount of available resources is an odd number, then the application of the Min-min algorithm produces better results. Otherwise, it would be better to start with max-min strategy. Max-min and Min-min are replaced instead to result in successive implementation of both, small and large cloudlets on different virtual machines and hence, the times waited for small cloudlets In Max-min and the large cloudlets in Min-min are overlooked. RASA does not contain any time-consuming instructions, the complexity of time within RASA is $0(mn^2)$ where m is the number of VMs and n is the number of cloudlets. RASA has weakness of both algorithms Max–min and Min–min, in spite of its manner to minimize them.

2.4. *Round Robin Allocation (RRA)*
In RRA [18][21][22],VMs (VM0, VM1, and VM2) was introduces in the system for the allocation strategy. According to this policy, cloudlet C0 allocated to VM0, C1 allocated to VM1, C2 allocated to VM2, C3 allocated to the VM0 and C4 allocated to VM1.
It distributes the cloudlets to the virtual machine that is available first, for instance, assume there are four cloudlets (C0, C1, C2, C3, C4) and three virtual machines {VM0, VM1, VM2} respectively available in the system. Based on this strategy, cloudlet C0 is allocated to VM0, C1 allocated to VM1, C2 to VM2, C3 is allocated to VM0 and C4 to VM1 respectively.
Round Robin Algorithm may assign large Cloudlets to the virtual machines with slowest processing speed thus, the processing time will get a long time to finish the execution that will Cause increasing the waiting time and the response time of the cloudlets as well. In addition, the highest virtual machine sometimes gets the smaller tasks that cause wasting in the resource utilization with reducing the whole performance.

2.5. *Conductance algorithm (CA)*
The CA algorithm [23] takes into consideration the size of every virtual machine as a pipe, the conductance is calculated using equation (1) as shown below, each virtual machine's processing speed ratio to the total of processing speeds of the virtual machines within the system. Dispensation speed of VMj is measured in million instructions per second (MIPS), presentation, however, is using$MIPSj$. Conductance of VMj is represented by $Conductance\ j$.

$$Conductance\ j = MIPSj / \sum_{j=1}^{n} MIPSj \quad (1)$$

Subsequent to the calculation of conductance, the number of cloudlets distributed to VMj is calculated by the multiplication of the $Conductance\ j$ of that specific VMj with the length of cloudlet list. In order to ensure the strip length of everyVMj, where strip length shall be represented by strip length, the use of equation (2) as shown below is implemented, it ensures the number of cloudlets processed by the virtual machine.

$$striplengthi = Conductance \times (length\ of\ cloudlet\ list) \quad (2)$$

Conductance Algorithm here might get wasted in the resource utilization due to the low VMs processing speed, which sometimes release rapidly and vice versa in the high VMs processing speed especially that the high VM has a length of the Cloudlets that is so large.

2.6 *Range wise Busy-checking 2-way Balanced (RB2B)*
It is a three-phase allocation algorithm [17], it is as follows:

(a) VM categorization phase
(b) Two round busy checking phase
(c) Cloudlet still not allocated (CSNA) phase.

A VM is considered appropriate only when it is not busy and it satisfies two-way balancing conditions. Thus, the process of RB2B is designated as follows:

The VMs are created and allocated to the host(s) and are reorganized in increasing order of processing speed (MIPS). A cloudlet arrives from the global queue (GQ) to the ADCB where the cloudlet allocation algorithm is executed. In the first phase, ADCB measures the length of the cloudlet and correspondingly chooses a VM following a cloudlet size acceptability range. If the chosen VM is available then ADCB will check it for a condition defined as Balance threshold. If the condition is satisfied the cloudlet will be allocated to the targeted VM. If the targeted VM is not available or the Balance threshold condition is not satisfied the ADCB will search for another VM which satisfies this condition. If such suitable VM is available, the cloudlet is allocated.

But if still no suitable VM is found then the third phase of RB2B will commence. In this phase ADCB will search for a VM according to EFT [24] which also satisfies the two-way balancing conditions i.e. Balance threshold and Local queue (LQ) length limitation, and the cloudlet will be queued to the local queue of this VM. There are not many restrictions of available algorithms and suggestions of the new algorithm: Min-min algorithm does not have uniform resource use in a way that makes it initiates by picking smaller cloudlets which take advantage of both virtual machines with more processing speed. Thus, the scheduling does not reach an optimum state, when the number of smaller cloudlets surpasses the bigger ones, to solve this issue, Max-min algorithm starts by scheduling the bigger cloudlets initially, however, sometimes, because of the execution of bigger cloudlets first, there can be an increase in the makespan. In addition to that, Max-min also increases smaller cloudlets' waiting time. RASA, on the other hand, has disadvantages from both sides, Min-min and Max-min, even though it rises to make them less. However, in RRA, big cloudlets are usually allocated to the virtual machines with low MIPS causing the delay in execution and the increase in the waiting time and response time within the cloudlets. Occasionally, it can occur that the most powerful virtual machines receive the smallest cloudlets making its resources use go waste, simultaneously depriving the overall performance, in the case of Conductance, low MIPS virtual machines occasionally achieve freedom quickly resorting to the waste of resources. The proposed algorithm RB2I performs in a way that the cloudlet will always be assigned to the most suitable virtual machine, based on the size of the cloudlet and the number of cloudlets distributed to the virtual machines is uniform to a precise extent. Making the percentage of resource allocation maximum and enhancing the completion time for every cloudlet so they become minimized when compared to other policies. This results in the minimization of the finish time, which brings us to the outcome of minimization of the makespan of the cloudlets.

3. Experiment Setup

The strategy proposed was simulated in Cloudsim, as illustrated in the figure 1. This model would interact with each other to act as a real system and each entity is described as below:

CIS (cloud information service) kind of registry that contains the resource available in the cloud

DC (a resource of the cloud) holds some hosts and it has some kind of hardware configurations, that can be a number of a processing element (PE it maybe (1,2,3)) and some RAM it might 512 MB, 1GB, 4GB in addition to the bandwidth.

The host might have the hardware configuration (Ram, BW and etc) and needs to register these characteristics to the CIS Broker. It needs to submit the task to DC and it has the characteristics of the DC from the CIS. Once the broker gets the characteristics of the DC and hosts, it interacts directly with the DC to assign the cloudlet to the VM.

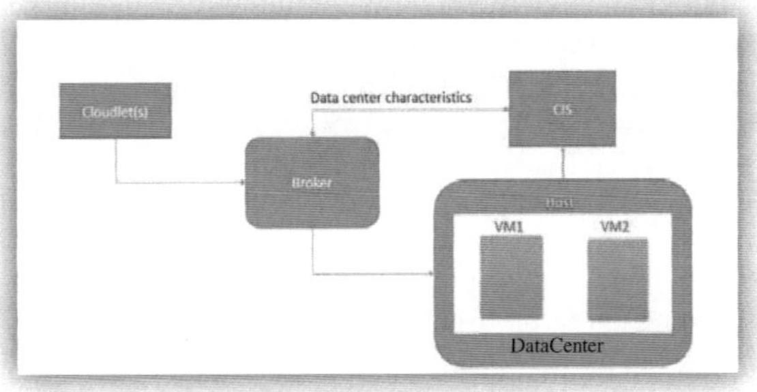

Figure 1. CloudSim model

4. Range wise Busy-checking 2-way Imbalanced

The proposed work is done in an imbalanced way based on the previous algorithm i.e. RB2B but, without considering the threshold balance and due to that, the work speeds up the makespan and keeps the VMs active.

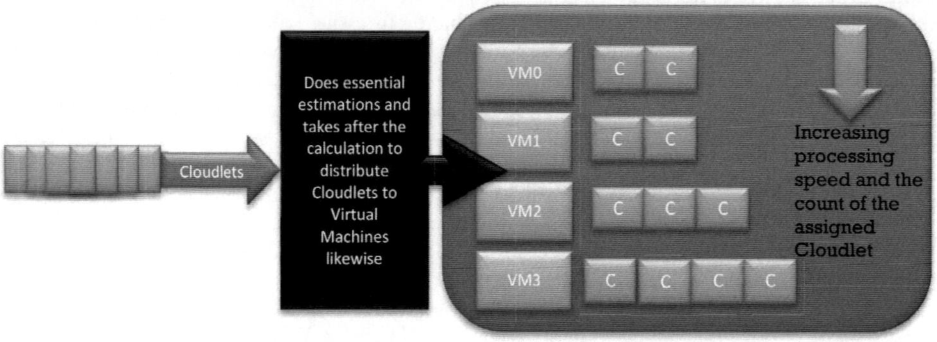

Figure 2. The proposed workflow model

As shown in figure (2), the VMs are sorted in the ascending manner, i.e. the VMi+1 is faster than VMi, that will be occupied by the cloudlets depending on the algorithm phases after getting calculated in the data center broker (e.g. previously determined). The whole process details will be shown below.

A three-phase allocation algorithm consists of three phases as the following:
(a) Primarily selected of the VMs,
(b) Two full round busyness checking,
(c) Cloudlet still not assigned.

The algorithm will be distributing Cloudlets among the VMs in a justice way. The process of the algorithm in a nutshell. Creating the VMs and assigning it to the Host(s) with classifying in increasing arrangement based on MIPS (the speed of the processing). Generating the Cloudlets and sending it to the Datacenter Broker (DCB) whereas the algorithm will get implemented. After that, the phases will initiate: Firstly, some equations will calculate in the DCB and determine the upper and lower limit for each VM to assign for it primarily if that VM was not busy. If the planned VM was unavailable (busy) for that specific cloudlet, then the second part of the algorithm will be running start.
In the second part, the cloudlet checks the VMs for availability, the last phase will begin. Datacenter Broker will go through which VM has least finish time of its task in addition to the specific Cloudlet, the least one will get the cloudlet in its local queue. The phases are described below in more detail.

4.1 Primarily selected of the VMs

The first step of the algorithm is to select the suitable VM to the cloudlets based on its length.
Thus at the beginning, the Data Center Broker (DCB) will be determined for each VM the acceptance range that illustrated below:
Determine the minimum and maximum cloudlet length (Cmax and Cmin):
Compute the total processing speed of the VMs that calls MIPS. As in equation (3) Total Processing Speed:

$$MIPStotal = \sum_{i=0}^{n=1} MIPSi \quad (3)$$

Compute the difference (X) between Cmax and Cmin over the total of MIPS as in equation (4), the different X,

$$X=(Cmax-Cmin)/MIPStotal \quad (4)$$

By assuming Fi is a MIPS of VMi and VMi are sorted in increasing order of the MIPS speed therefore MIPSi+1>MIPSi. Thus, the distribution of the processing speed range will demonstrate in figure 3.

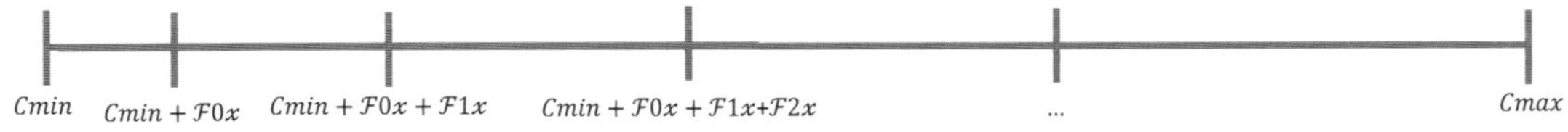

Figure 3. Range distributions among VMs

Compute the upper and lower limits of Cloudlets length that accepted by each VM by a linear sequential equation as per equation (5) and equation (6) respectively:

$$Cmin + \mathcal{F}0x + \mathcal{F}1x \ldots + \mathcal{F}m^{x} \quad (5)$$

$$Cmin + \mathcal{F}0x + \mathcal{F}1x + \ldots + \mathcal{F}m - 1X + 1 \quad (6)$$

Table(1) below shows the acceptability range equations:

Table 1. The acceptability ranges of the cloudlet length

#VM	Parameters	Number of parameters
VM_0	$Cmin$	$Cmin + \mathcal{F}0x$
VM_1	$Cmin + \mathcal{F}0x$+1	$Cmin + \mathcal{F}0x + \mathcal{F}1x$
VM_m	$Cmin + \mathcal{F}0x + \mathcal{F}1x + \ldots + \mathcal{F}m - 1X + 1$	$Cmin + \mathcal{F}0x + \mathcal{F}1x + \ldots + \mathcal{F}m^{x}$
VM_{n-1}	$Cmin + \mathcal{F}0x + \mathcal{F}1x + \ldots + \mathcal{F}n - 2X + 1$	$Cmin + \mathcal{F}0x + \mathcal{F}1x + \ldots + \mathcal{F}n - 1 = Cmax$

At the end of this phase, the DCB will target for each cloudlet a suitable VM based on the lower and upper limit and term it as targetedVM.

4.2 Two round busy checking phase

This phase will initiate once the Cloudlet finds the targetedVM is busy and not available at the time of the cloudlet arrival thus, two round of busyness checking will start:

The first round the DCB will check the next higher processing speed VM if it is busy or not. This process will continue until the cloudlet reaches to the highest processing speed VM, if all higher VMs are busy then the second round will begin. In second round, the DCB will check the next lower processing speed VMs, if it is busy or not. As illustrated in Figure 4.

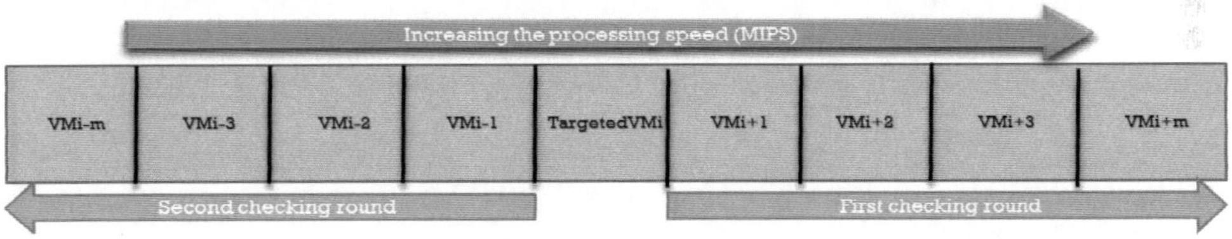

Figure 4. Two round phase.

In this phase, the DCB will check the higher MIPS VMs one by one until finding a suitable VM to the cloudlet. If the DCB could not find the suitable VM, then the second round of this phase will initiate.

The DCB will do same manner by trying to find any suitable VM starting from the higher to the lower MIPS VMs. If it is found, then the DCB will submit the cloudlet to the suitable and available VM. Otherwise, phase three will commence.

4.3 Cloudlet-still-not-allocated phase (CSNA)

In third phase of the algorithm , in case the cloudlet did not submit during the previous phases, then it will move to this phase to get submission by calculating the finish time of that cloudlet on all VMs, in addition to the waiting time for the VMs and make comparisons among the result of that to see which VMs will give least finish time to submit the cloudlet in its local queue.

RB2I Algorithm pseudo code

```
class RB2I {
  set The MIPS speeds
   calculate the sum of MIPS
   set parameters for the VM
   for i=0 to no of VMs
    {
     Create VM[i] //VMi<VMi+1
    }
   Set parameters for Cloudlets
   for i=0 to no of Cloudlets
    {
     Create hetrogonous Cloudlets
    }
 DataCenterBrocker Class {
   Determind the upper & lower limits for the Cloudlet
   PA=primary assign the VM to the Cloudlet
   Busy=vm.getTotalUtilizationOfCpu(CloudSim.clock())
   While (i<Cloudlet.size)
    {
     if (PA!=Busy)
       submit the cloudlet
     if (PA==Busy)
       if (Cloudlet == upperlimit)
         {
         Check the VM next slower MIPS if not Busy then submit to it
         if (VMs==checked)
           {
           Go to Phase 3
           }
        else
        CPA=current PA
         for j=CPA+1 to no of VM
         if (VM[j] != Busy)
         {
          submit to VM[j]
         }
         for j=CPA-1 to VM.0
          {
          if (VM[j] != Busy)
            submit to VM[j]
          }
    if (VMs==checked)&(PA==Busy)
           {
           Go to Phase 3
           }
 phase 3
 {
 for (i=0 to no of VMs)
  {
   FT[i]=finish time for the VM[i]
   FTS[i]=finish time for the specific task on that VM
   C[j]=FT[i]+FTS[i]
  }
 compute the min C[j] time
submit  }
```

4. Result and discussion

The result shows the performance evaluation for the proposed work with the algorithm RB2B and the comparison between them. Due to the difficulties in the testing, handling and assessing on a real system, this research procedure has been done on the discrete event simulator called Cloudsim to emulate the actual environment. The main parameters applied to assess the proposed algorithm are shown in table (2).

Table 2. Simulation Parameters for RB2I

Parameters	Number of parameters
Number of Cloudlet	1000
Cloudlet length	(1000-100,000) MI
Number of VMs	10
MIPS for VM	(1000-10,000) MIPS
Number of DataCenters	1
Number of Hosts	2
pesNumber (Number of CPUs)	1
num_user (Number of users)	1
Host(s) Storage	1,000,000 MB
Host(s) memory	4096 MB

The makespan is the maximum finish time (FTi) among all received tasks per time unit. Or in other words, it is the last finish time to the specific VM per time unit. It measures the proficiency and the intelligence of the way of the scheduling algorithm in mapping the tasks to the cloud suppliers. From mapping proficiency point of view, this parameter considers the reflector for the behavior of algorithms. Figure (5) depicts this parameter. It shows a set of 1000 cloudlets with length ranging from 1000 to 100000 MI (Million instruction) by using heterogeneous manner and set of 10 VMs with different processing speed ranging from (1000-10,000) MIPS.

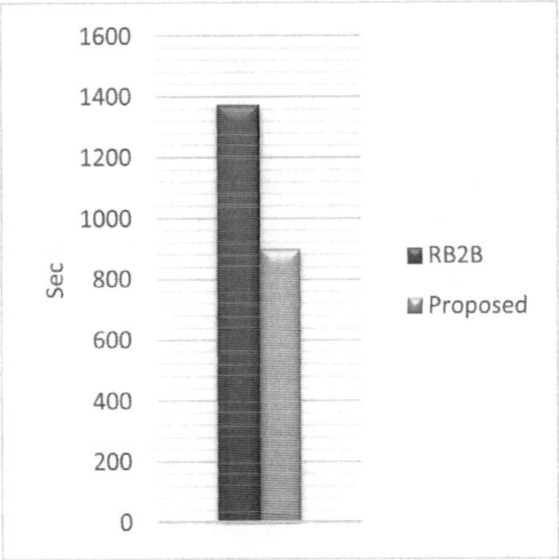

Figure 5. Makespan Time

The proposed work showed an enhancements of 34.53% in the average makespan time compared with the balanced algorithm (RB2B).

5. Conclusion

The cloud computing is a massive field of study hence the Cloudlet allocation takes a vital role in delivery the service in an appropriate way. Thus, there is an immense scope of improvement in this field, this work presents a three stages of cloudlet allocation algorithm that overcomes the existing allocation strategies weaknesses in a proficiently manner where in the three stages of RB2I tried to allocate the cloudlet in the suitable VM, hence in the first stage measures the length for each Cloudlet and accordingly elects VM pursuing a cloudlet size acceptability limit. If the condition is satisfied, the cloudlet will be allocated to the targeted VM. If such suitable

IOP Conf. Series: Journal of Physics: Conf. Series **1018** (2018) 012018 doi:10.1088/1742-6596/1018/1/012018

VM is available i.e. it will check the higher VMs by comparing the low VMs, once the cloudlet find the suitable and available VM, the cloudlet is allocated. However if it is still not suitable and available (all VMs are Busy), then the third stage of RB2I will initiate. Then, it examines for an appropriate VM by computing the consuming time to finish the cloudlet execution. Then it allocates the cloudlet to the local queue of choosen VM. Therefore, the objective of cloudlet allocation and well resources manipulation could be succeeded in the less average Makespan time.

References

[1] Fu X and Zhou C 2015 Virtual machine selection and placement for dynamic consolidation in Cloud computing environment *Front. Comput. Sci.* **9** 322

[2] Zhang Y, Koelbel C and Kennedy K 2007 Relative performance of scheduling algorithms in grid environments. *In Proceedings of the Seventh IEEE international Symposium on Cluster Computing and the Grid. CCGRID. IEEE Computer Society, Washington, DC* 521–8

[3] Radulescu A, Van Gemund A 1999 On the complexity of list scheduling algorithms for distributed memory systems *In: Proceedings of the 13th International Conference on Supercomputing ICS '99. ACM, New York, NY* 68–75

[4] Carino R and Banicescu I 2008 Dynamic load balancing with adaptive factoring methods in scientific applications. J. Supercomput. **44** 41–63

[5] Rashid M, Banicescu I and Carino R L 2008 Investigating a dynamic loop scheduling with reinforcement learning approach to load balancing in scientific applications *In: Proceedings of the 2008 international Symposium on Parallel and Distributed Computing (July 01- 05, 2008). ISPDC. IEEE Computer Society, Washington, DC* 123–30

[6] Aziz A and El-Rewini H 2008 On the use of meta-heuristics to increase the efficiency of online grid workflow scheduling algorithms *Cluster Comput* **11** 373–90

[7] Pop F 2008 Communication model for decentralized meta-scheduler in Grid Environments *In: Proceedings of The Second International Conference on Complex, Intelligent and Software Intensive System, Second International Workshop on P2P, Parallel, Grid and Internet computing – 3PGIC-008 (CISIS'08). IEEE Computer Society, Barcelona, Spain, ISBN 0-7695-3109-1* 315–20

[8] Braun T D, Siegel H J, Maciejewski A A and Hong Y 2008 Static resource allocation for heterogeneous computing environments with cloudlets having dependencies, priorities, deadlines, and multiple versions *J. Parallel Distrib. Comput* **68** 1504–16

[9] Makelainen M, Khan Z, Hanninen T and Saarnisaari H 2014 Algorithms for opportunistic load balancing cognitive engine. *In: Cognitive Radio Oriented Wireless Networks and Communications (CROWNCOM), 2014 9th International Conference, 2-4 June 2014* 125–30

[10] George Amalarethinam D I and Muthulakshmi P 2011 An Overview of the scheduling policies and algorithms in Grid Computing *Int. J. Res. Rev. Comput. Sci* **2** 280–94

[11] Dong F, Luo J, Gao L and Ge L 2006. A grid task scheduling algorithm based on QoS priority grouping *The Proceedings of the Fifth International Conference on Grid and Cooperative Computing (GCC'06). IEEE* 58–61

[12] Khanli L M and Analoui M 2008 Resource scheduling in desktop grid by grid-JQA *The 3rd International Conference on Grid and Pervasive Computing. IEEE* 25–8

[13] Khanli L M and Analoui M 2007 Grid_JQA: a QoS guided scheduling algorithm for grid computing *The Sixth International Symposium on Parallel and Distributed Computing (ISPDC'07). IEEE* 34

[14] Elmroth E and Tordsson J 2008 Grid resource brokering algorithms enabling advance reservations and resource selection based on performance predictions *J. Future Gener. Comput. Syst.* **24** 585–93

[15] Benjamin Khoo B T, Veeravalli B, Hung T and Simon See C W 2007 A multi-dimensional scheduling scheme a Grid computing environment *J. Parallel Distrib. Comput* **67** 659–73

[16] Livny M and Melman M 2011 Load balancing in homogenous broadcast distributed systems *Proceedings of the ACM Computer Network: Performance Symposium* 47–55

[17] Roy S, Banerjee S, Chowdhury K R and Biswas U 2016 Development and analysis of a three phase cloudlet allocation algorithm *J. King Saud Univ. Comput. Inf. Sci*

[18] Parsa S and Entezari-Maleki R 2009 RASA: a new grid cloudlet scheduling algorithm *World Appl. Sci. J* **7** 152–60

[19] Kumar P and Dutta Pramanik P K 2012 Host selection methodology in cloud computing environment *Int. J. Adv. Res. Comput. Eng. Technol. (IJARCET)* **1** 2278-1323

[20] El-kenawy E T, El-Desoky A I and Al-rahamawy M F 2012 Extended Max–min scheduling using petri net and load balancing *Int. J. Soft Comput. Eng. (IJSCE)* **2** 2231-307

[21] Bhatia W, Buyya R and Ranjan R 2010 CloudAnalyst: a CloudSim based visual modeler for analysing cloud computing environments and applications, *24th IEEE International Conference on Advanced Information Networking and Applications* 446–52

[22] Banerjee S, Adhikari M, Kar S and Biswas U 2015 Development and analysis of a new cloudlet allocation strategy for QoS improvement in cloud *Arabian J. Sci. Eng* **40** 1409–25

[23] Chatterjee T, Ojha V K, Adhikari M, Banerjee S, Biswas U and Snasel V 2014 Design and implementation of a new datacenter broker algorithm to improve the QoS of a cloud. *In: Proceedings of ICBIA 2014, Advances in Intelligent Systems and Computing, 303.Springer International Publishing, Switzerland* 281–90

[24] Bittencourt L F, Sakellariou R and Madeira E R M 2010 DAG scheduling using a look ahead variant of the heterogeneous earliest finish time algorithm *In: 18th Euromicro International Conference on Parallel, and Network-Based Processing (PDP)* **34** 27

IOP Conf. Series: Journal of Physics: Conf. Series **1018** (2018) 012019 doi:10.1088/1742-6596/1018/1/012019

A Review on Data Stream Classification

Haneen A. A[1], A. Noraziah[1, 2], Mohd Helmy Abd Wahab[3]

[1]Faculty of Computer Systems & Software Engineering, Universiti Malaysia Pahang, 26300 Kuantan, Pahang.
[2]IBM Centre of Excellence, Universiti Malaysia Pahang, 26300 Kuantan, Pahang.
[3]Department of Computer Engineering, Faculty of Electrical and Electronic Engineering
Universiti Tun Hussein Onn Malaysia, 86400 Batu Pahat, Johor, Malaysia
Email: noraziah@ump.edu.my

Abstract. At this present time, the significance of data streams cannot be denied as many researchers have placed their focus on the research areas of databases, statistics, and computer science. In fact, data streams refer to some data points sequences that are found in order with the potential to be non-binding, which is generated from the process of generating information in a manner that is not stationary. As such the typical tasks of searching data have been linked to streams of data that are inclusive of clustering, classification, and repeated mining of pattern. This paper presents several data stream clustering approaches, which are based on density, besides attempting to comprehend the function of the related algorithms; both semi-supervised and active learning, along with reviews of a number of recent studies.

Keywords: Clustering, Data Mining, Data Streams, Computational Intelligence

1. Introduction

The dramatic growth in information technology and the vast volume of generated data have led to several fresh challenging discovery tasks in data processing. In fact, the notion "data stream" refers to the data sequence that is embedded continuously and uncertainly in a certain system for storage or for processing[1]. Moreover, these data streams share similar characteristics from the aspects of massiveness, temporal order, rapid changes, as well as potential length infinite [2-4].

In addition, Aggarwal et al. has offered some reasons that disport data streams from the traditional data mining approach[5] which are: (i) size of data streams is potentially boundless; (ii) elements of stream arriving on-line; (iii) limitations in memory space, prior to processing an element, and system rejection or summary; (4) failure of system in controlling or determining the arrival of data elements.

A number of algorithms have been opted relevant to this study. Nevertheless, these selected algorithms fail when clustering data with varied densities due to usage of common parameters meant for clustering that is based on density, thus generating poor setting for densities that are of single and multiple types. Clustering algorithms with density-based data stream.

2. Data stream clustering

Clustering, the task of information focuses to (ordinarily *k*) groups to such an extent that focuses inside each groups are more like each other than to focuses in different groups, is an extremely essential unsupervised information mining assignment. For static informational collections, strategies like *k*-means, *k*-medoids, hierarchical clustering and density-based techniques have been produced among others. A number of these methods are available in tools like *R*. However, the standard algorithms require access to all information and all data points and typically iterate over the data multiple times. This prerequisite makes these algorithms unsuitable for large data streams and led to the development of data stream clustering algorithms. Figure 1 shows the high level view of the stream architecture.

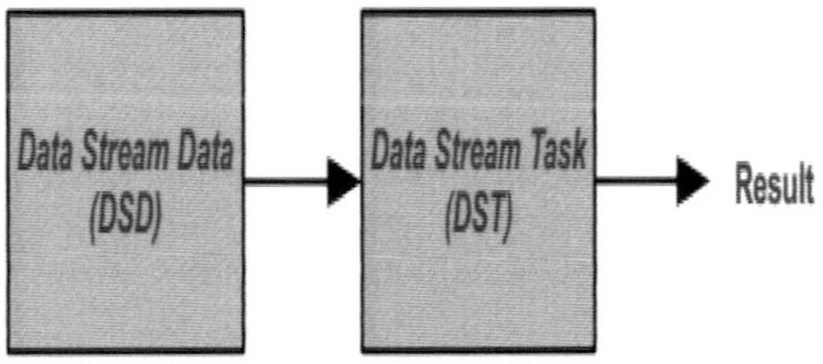

Fig.1. High level view of the stream architecture

Many algorithms for data streams over the last 10 years have been proposed for clustering. Most data stream clustering algorithms deal with the problems of unbounded stream size, and the requirements for real-time processing in a single pass by using the following two-stage online/offline approach:

1. Online: Summarize the information by a set of k' micro-clusters groups composed in a space efficient information structure which also enables fast look-up. Micro-clusters were introduced for CluStream[3] based on the idea of cluster features developed for clustering large data sets with the BIRCH algorithm. Micro-clusters are representatives for sets of similar data points and are created using a single pass over the data (normally progressively when the information of data stream arrives). Micro-clusters are frequently represented by cluster centres and additional statistics such as weight (local density) and dispersion (variance). Each new data point is assigned to its closest (in terms of a similarity function) micro-cluster. Some algorithms use a grid instead and micro-clusters are represented by non-empty grid cells [4]. If a new data point cannot be assigned to an existing micro-cluster, a new micro-cluster is created. The algorithm might also perform some housekeeping (merging or deleting micro-clusters) to keep the number of micro-clusters at a manageable size or to remove information outdated due to a change in the stream's data generating process [5].
2. Offline: When the user or the application requires a clustering, the k' micro-clusters are clustered into k « k' final clusters sometimes referred to as macro-clusters. Since the offline part is typically not respected time critical, most researchers utilize a regular clustering algorithm where micro-cluster centers are viewed as pseudo-points. Average clustering techniques involve k-means or clustering based on the concept of reachability introduced by DBSCAN. This algorithms is often modified to take also the weight of micro-clusters into account.

The most popular approach to adapt to concept drift (changes of the data generating process over time) is to use the exponential fading strategy introduced first for DenStream by [4]. Micro-cluster weights are faded in every time step by a factor of $2-\lambda$, where $\lambda > 0$ is a user-specified fading factor. This way, new data points have more impact on the clustering and the influence of older points gradually disappears. Alternative models use sliding or landmark windows. Details of these methods as well as other data stream clustering algorithms are discussed in[4].

Clustering algorithms that are based on density are meant to identify random-shaped clusters, as these clusters are determined by looking into the areas of density. For example, when a couple of points are relatively near at a dense area, these points could link to generate one cluster. Some instances of this approach are DBSCAN [6], OPTIC[7], and DENCLUE [8]. Moreover, such methods have been employed within these recent times to cluster data streams that are evolving[9-11].

Besides, the DenStream[12] has extended the notion of micro-cluster [13], besides introducing both the potential and the outlier for these micro-clusters in order to differentiate the actual data from the ones that are not. In fact, this technique embeds the stages of online and offline. As for the online stage at the initiation phase, the DBSCAN algorithm is employed by the DenStream for the few points that are initial, thus forming potentially the first micro-clusters. Nonetheless, any new point of data is attached to the closest micro-cluster with potential or even to a micro-cluster that is an outlier. However, upon failure of attachment, a fresh micro-cluster that is outlier in nature is built and located at the outlier region. On the other hand, the DBSCAN is used in the offline stage in order to identify clusters that are final found on micro-clusters with potential. Furthermore, DenStream promotes a method of pruning to determine the weight of micro-clusters that are outlier found at the outlier region. Unfortunately, this technique fails in clustering data stream that has varied densities because density is dismissed when forming micro-clusters at the online stage result is in Figure 2.

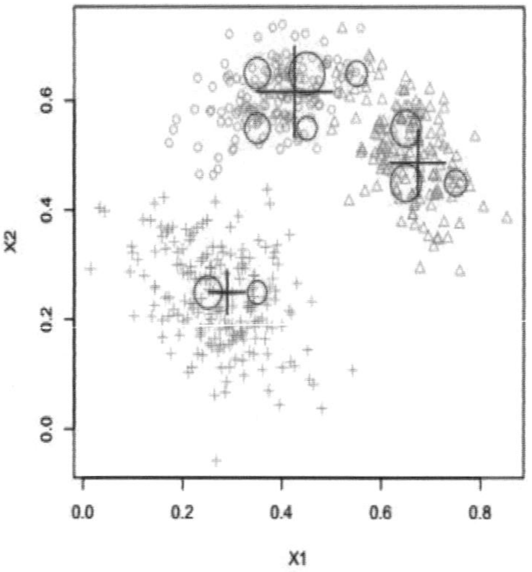

Fig.2. Data stream clustering result of D-stream on a simple simulated data set.

Meanwhile, the offline stage also leads to the formation of final clusters with the use of DBSCAN[14] that fails in determining data with density. In addition, the process of attaching takes up a lot of time. On the other hand, the FlockStream [15] refers to a type of clustering algorithm for density and it adopts the bio-inspired model, which reflects the model of flocking [16], where micro-clusters function as independent agents that generate formation of clusters. In fact, an agent is considered for the points of data via mapping on virtual space. Furthermore, these agents levitate for a certain time within a predetermined range of visibility. However, upon visiting an agent with similar attributes, cluster is formed, thus generating the stages of online and offline for cluster formation occurs without prediction.

Additionally, a clustering framework that is based on density grid is suggested by[17] meant for real time data streams, or also known as D-Stream I, which possesses both stages of online and offline. At the online stage, a fresh point of data is read, mapped onto a grid, and lastly, the grid attribute vector is updated. In oppose, the clusters are modified at offline stage at interval time. This gap of interval time is reflective of the minimal transformation time for varied grids. Besides, this D-Stream I initially keeps the grid density updated, and later, uses a common technique to cluster based on density. Significantly, this particular model adopts outliers are grids of sporadic, which refers to a grid that is sparse with limited data and fails converting to a grid that is dense.

On top of that, D-Stream I lowers the threshold of density in accordance to its function. Hence, a grid is known as sporadic if the density of the sparse grid is lower that the threshold. Besides, the pruning stage occurs at interval time, where modification of clusters and removal of sporadic grid occur. Furthermore, a hash table is used in D-Stream I to retain the list of grids. Next, D-Stream I was expanded [18], thus called as D-Stream II, which disregards the position of data in a grid for clustering algorithm with multi-density.

On the other hand, [19] developed MR-Stream, which is a clustering algorithm based on density meant for streams of data with numerous resolutions[20] mainly to enhance its function by ensuring that the offline phase is continuously run because this algorithm projects the accurate time for clusters generation. This works by storing space partitioning by providing partition of space data in cells, as well as data structure that resembles a tree, which stores clustering data in varied resolutions. Hence, every node possesses data concerning its parent and children. Furthermore, the MR-Stream works based on online and offline stages.

At the stage of online, the fresh and newly arrived point of data is mapped on the grid. Meanwhile, if there is no sub-node at the tree structure, a new one is developed to store detail regarding the parent until the tree root. Besides, pruning is done at interval. Next, clusters are developed based on that defined by users at the offline stage to mark reachable cells that are dense at a certain distance as a cluster. As such, clusters that are noisy are eliminated after comparing size and densities with those of the thresholds. Moreover, the MR-Stream adopts the method of memory sampling to identify the exact time to run the offline element, thus enhancing its performance.

Meanwhile, HDC-Stream [21] refers to a clustering algorithm based on density for data stream applied for Internet of Things (IoT). Nonetheless, this algorithm projects processing time that is low, hence appropriate devices of IoT with real-time usage. Besides, clustering is generated by its hybrid approach with three steps, under stages of online and online. The HDC-Stream online stage reads fresh data in a continuous manner, thus adding them into grid or mini-cluster. As for outliers, HDC-Stream prunes them often. Meanwhile, the final clusters are generated at the offline stage, based on users. In fact, this method is high in quality, but has intricacy at lower time for data stream clustering. Nevertheless, its drawback is that it cannot cluster data with multiple densities.

Besides, a comprehensive study about clustering algorithms with density was conducted by dismissing the offline clustering stage [22]. In fact, comparison of new point via FlockStream with other aspects within the visibility distance has decreased neighbourhood comparison for points. This distance of visibility refers to the threshold determined by user. Hence, the related algorithm is a failure for data with multiple densities that possess high quality due to its single distance.

Moreover, each agent has to adhere to some regulations over time to get located within the virtual space, for instance, separation, alignment, and cohesion [23]. With that, there are three types of agents in FlockStream: fresh points of data have fundamental representative agents, while potential and outlier micro-clusters are represented by p- and o-representative agents. In precise, a representative agent is generated as agents with similar attributes combine, which relies on their weights. Nonetheless, FlockStream possesses higher time for computation and decreases the amount of comparisons, when compared to DenStream.Although the procedure for clustering is similar for both D-Stream and D-Stream I; correlation could occur between grids that are dense in D-Stream II, whereby these grids are deemed as strongly correlated if they have higher attraction between grids, in comparison to that of predetermined threshold. Furthermore, the list of grids is kept in tree, and not table, to enable rapid processing and to save memory space.

3. Clustering algorithms with multi-density data set

The GMDBSCAN[24] refers to a clustering algorithm for multiple densities that employs the grid method. In fact, this algorithm identifies the local parameters of MinPts through the use of density grid, which are clustered via DBSCAN by the selected MinPts local parameters. Besides, similarity in density determines the integration among this two-pass clustering; therefore inapt for use of data streams. On the other hand, MSDBSCAN [25] refers to a clustering algorithm based on density in the light of the notion of essential points found in DBSCAN, depending on the location of the neighbours. Furthermore, this suggests a new approach for the essential points, which are known as local core distance (lcd), to reflect the MinPts objects distance. Besides, MSDBSCAN determines the value of lcd for the available points of data to identify the essential, which shares similar value with the lcd vector. Moreover, these essential points are used by the algorithm to generate huge clusters.

Nevertheless, MSDBSCAN has high time intricacy, thus unsuitable for streams of data. Generally, this multiple DBSCAN[26] blankets the issue related to multiple density, which is determined by various values of ϵ. Thus, "must link constraint" and "kth" are employed at the nearest distance to obtain values for ϵ at varied densities. Besides, the multiple DBSCAN selects the best value of ϵ for distribution of density by applying the algorithm that detects outlier. Next, DBSCAN is carried out based on the data by determining the values of ϵ.

In another case, [27] developed a clustering algorithm that is hierarchical to generate clusters with multiple density. This algorithm applies an approach that has multiple levels in order to identify clustered structures that are hierarchical from the data. As such, the agglomerative k-means is employed to build a tree for cluster with varied densities. Moreover, this approach of cluster validation can be employed to determine both clusters of composite and atomic. For instance, clusters that are atomic disregards further division. Meanwhile, division is performed using agglomerative k-means for clusters that are composite, where the repeated process generates the cluster tree, which is also reflective of a clustering algorithm with dual-pass.

Moving on, the IS-DBSCAN [28] refers to an approach that enhances the algorithm of DBSCAN for limited parameters, as well as its ability in clustering data with multiple densities. This approach introduces a fresh notion called 'space ranking', where the points of data are ranked within space depending on metrics density. On top of that, the ISDBSCAN applies the concept of Influence Space (IS) initiated in INFLO (INFLuenced Outlierness)[27]. Basically, IS offers more accurate prediction for the distribution of neighbourhood density, besides enhancing clusters division with varied densities. Moreover, the conventional technique of determining ϵ-neighbourhood is substituted by a technique that applies IS meant for clustering based on density.

Meanwhile, SCDM2 [21] refers to the extension of SCDM[22], thus a clustering algorithm that is semi-supervise. The SCDM2 was meant for data with multiple density for it applies constriction in the process of clustering. Moreover, in SCDM, the parameter ϵ in DBSCAN is determined based on "must link constraint". Hence, when a cluster does not have any constriction, it would not emerge in the final cluster. As for SCDM2, those clusters that cannot be linked are also embedded to the ϵ representative. On top of that, the DBSCAN-DLP (multiple density DBSCAN based on Density Levels Partitioning)[28] ensures that the parameters used for every cluster to identify the multiple densities clusters via division of density level. Initially, the data sets are separated into various levels of densities. After that, the value of ϵ is calculated for every level of density.

Finally, the DBSCAN is employed to carry out clustering for every level of density, along with values of ϵ to obtain results of clustering. In fact, the DBSCAN-DLP is reflective of clustering algorithm with dual-pass that possesses computation time that is rather high to be used for streams of data. Other than that, GDCLU[24] refers to a clustering algorithm based on density employing the grid approach. Besides, this algorithm, which is scale independent, suggests that this grid is density-based in relation to the neighbours.

Next, the DSCLU[29] employs the approach of micro-clustering for data stream with multiple density. This particular algorithm identifies the micro-clusters that are dominant by taking weights of neighbours into consideration. Hence, these micro-clusters become dense, in fact, share similar density with their neighbours. As for the offline stage, the dominant micro-clusters are used for clustering. Besides, this technique employs the same radius for micro-clusters formation.

4. Conclusions

This particular study has looked into several clustering algorithms that are based on single and multi-density meant for data stream. Furthermore, these grid-based algorithms have their points of data marked on grids, which undergo systematic process. Besides, the values of density for the points of data indicate if the grids are sparse or dense. In fact, structures with grid help in addressing drawback related to algorithm with legacy partition, which is inapt when used for data streams. Moreover, density enhances clustering to be more rapid and more reliable. Thus, issues related to clustering of data stream are investigated, thus seeking solutions to these glitches. The following are the main issues found within this area: Constraint in Memory: Memory has emerged as a huge limitation due to the huge data stream size. Since it is not possible to keep all data in memory, performing clustering is rather tricky. Therefore, a portion of the data is placed in a time window for clustering purpose, in which the data are later passed to free memory, and lastly, moved forward.

Constraint in Time: Time is also a hindrance for clustering of data stream because rapid stream and data with continuous flow do not wait for the processing. Therefore, clustering is not only done within a certain time, but the system has to continue accepting data without any problem. Hence, one solution to this is the enhancement of algorithm in both response time and efficiency in handling data with a rapid pace.

Acknowledgments

Appreciation conveyed to Ministry of Higher Education, Malaysia for financial assistance under Fundamental Research Grant Scheme RDU140101. In addition, our appreciation convey to the Universiti Malaysia Pahang for project financing under Research Grant Scheme RDU170398 and PGRS170304.

References

[1]. Aggarwal, C.C., *Data streams: models and algorithms*. Vol. 31. 2007: Springer Science & Business Media.

[2]. Rajaraman, A. and J.D. Ullman, *Mining of massive datasets. 2012.* Cited on, 2011: p. 139.

[3]. Kholghi, M., H. Hassanzadeh, and M. Keyvanpour. *Classification and evaluation of data mining techniques for data stream requirements*. in *Computer Communication Control and Automation (3CA), 2010 International Symposium on*. 2010. IEEE.

[4]. Gama, J., *Knowledge discovery from data streams*. 2010: CRC Press.

[5]. Aggarwal, C.C., et al. *A framework for clustering evolving data streams*. in *Proceedings of the 29th international conference on Very large data bases-Volume 29*. 2003. VLDB Endowment.

[6]. Aggarwal, C.C. and C.K. Reddy, *Data clustering: algorithms and applications*. 2013: CRC press.

[7]. Amini, A., H. Saboohi, and T.Y. Wah. *A multi density-based clustering algorithm for data stream with noise*. in *Data Mining Workshops (ICDMW), 2013 IEEE 13th International Conference on*. 2013. IEEE.

[8]. Amini, A., et al., *A fast density-based clustering algorithm for real-time internet of things stream*. The Scientific World Journal, 2014. 2014.

[9]. Amini, A. and T.Y. Wah. *Density micro-clustering algorithms on data streams: A review*. in *Proceeding of the International Multiconference of Engineers and Computer scientists (IMECS)*. 2011.

[10]. Amini, A., T.Y. Wah, and H. Saboohi, *On density-based data streams clustering algorithms: a survey*. Journal of Computer Science and Technology, 2014. 29(1): p. 116-141.

[11]. Amini, A., et al. *A study of density-grid based clustering algorithms on data streams*. in *Fuzzy Systems and Knowledge Discovery (FSKD), 2011 Eighth International Conference on*. 2011. IEEE.

[12]. Ankerst, M., et al. *OPTICS: ordering points to identify the clustering structure*. in *ACM Sigmod record*. 1999. ACM.

[13]. Cao, F., et al. *Density-based clustering over an evolving data stream with noise*. in *Proceedings of the 2006 SIAM international conference on data mining*. 2006. SIAM.

[14]. Cassisi, C., et al., *Enhancing density-based clustering: Parameter reduction and outlier detection*. Information Systems, 2013. 38(3): p. 317-330.

[15]. Chen, L., L.-J. Zou, and L. Tu, *A clustering algorithm for multiple data streams based on*

spectral component similarity. Information Sciences, 2012. 183(1): p. 35-47.

[16]. Chen, X., et al., *An improved semi-supervised clustering algorithm for multi-density datasets with fewer constraints.* Procedia Engineering, 2012. 29: p. 4325-4329.

[17]. Chen, Y. and L. Tu. *Density-based clustering for real-time stream data.* in *Proceedings of the 13th ACM SIGKDD international conference on Knowledge discovery and data mining*. 2007. ACM.

[18]. Esfandani, G. and H. Abolhassani, *MSDBSCAN: multi-density scale-independent clustering algorithm based on dbscan.* Advanced Data Mining and Applications, 2010: p. 202-213.

[19]. Nakata, Y., T. Hochin, and H. Nomiya, *2012 13th ACIS International Conference on Software Engineering, Artificial Intelligence, Networking and Parallel & Distributed Computing (SNPD).*

[20]. Boden, B., M. Ester, and T. Seidl. *Density-based subspace clustering in heterogeneous networks.* in *Joint European Conference on Machine Learning and Knowledge Discovery in Databases*. 2014. Springer.

[21]. Forestiero, A., C. Pizzuti, and G. Spezzano, *A single pass algorithm for clustering evolving data streams based on swarm intelligence.* Data Mining and Knowledge Discovery, 2013: p. 1-26.

[22]. Huang, T.-q., et al. *Reckon the parameter of dbscan for multi-density data sets with constraints.* in *Artificial Intelligence and Computational Intelligence, 2009. AICI'09. International Conference on*. 2009. IEEE.

[23]. Li, X., et al., *On cluster tree for nested and multi-density data clustering.* Pattern Recognition, 2010. 43(9): p. 3130-3143.

[24]. Namadchian, A. and G. Esfandani. *DSCLU: a new Data Stream CLUstring algorithm for multi density environments.* in *Software Engineering, Artificial Intelligence, Networking and Parallel & Distributed Computing (SNPD), 2012 13th ACIS International Conference on*. 2012. IEEE.

[25]. Wan, L., et al., *Density-based clustering of data streams at multiple resolutions.* ACM Transactions on Knowledge discovery from Data (TKDD), 2009. 3(3): p. 14.

[26]. Xiong, Z., et al., *Multi-density dbscan algorithm based on density levels partitioning.* JOURNAL OF INFORMATION &COMPUTATIONAL SCIENCE, 2012. 9(10): p. 2739-2749.

[27]. Hu, W., et al. *Research on Parallel Data Stream Clustering Algorithm Based on Grid and Density.* in *Computer Science and Mechanical Automation (CSMA), 2015 International Conference on*. 2015. IEEE.

[28]. Sun, Z., et al. *Knowledge-based evolving clustering algorithm for data stream.* in *Service Systems and Service Management (ICSSSM), 2014 11th International Conference on*. 2014. IEEE.

[29]. Hahsler, M. and M. Bolaños, *Clustering Data Streams Based on Shared Density Between Micro-Clusters.* IEEE Transactions on Knowledge and Data Engineering, 2016. 28(6): p. 1449-1461.

IOP Conf. Series: Journal of Physics: Conf. Series **1018** (2018) 012021 doi:10.1088/1742-6596/1018/1/012021

Two Stages Transfer Algorithm (TSTT) for Independent Tasks Scheduling in Heterogeneous Computing Systems

[1]Abdulrahman K. Al-Qadhi, [1]Ahmad Alauddin Ariffin, [1,3] Rohaya Latip, [1]Nor Asila Wati Abdul Hamid, [2]Ammar S. Al-Zubaidi

[1]Universiti Putra Malaysia, Faculty of Computer Science and Information Technology, Selangor, Malaysia
[2]University of Baghdad, Computer Centre, Baghdad, Iraq
[3]Institut Penyelidikan Matematik (INSPEM), Universiti Putra Malaysia, Selangor, Malaysia
E-mail: prog.abdulrahman@gmail.com

Abstract. Task scheduling is critical in heterogeneous systems especially with the huge number of tasks transmitted over grid causing system delay. Since heuristics are proposing methods for solving heterogeneous computing systems, several techniques were proposed for the scheduling on grid computing systems to get better execution time. In this paper, a proposed new heuristic algorithm named Two Stages Tasks Transfer (TSTT) algorithm to enhance Tenacious Penalty Based scheduling (TPB) algorithm. Heterogeneous Computing Scheduling Problem (HCSP) mathematical model has been used, where the independent tasks assigned to heterogeneous machines with different characteristics. Twelve datasets with different heterogeneity level examined using different heuristic algorithms to compare the performance with our proposed algorithm. The proposed algorithm showed its efficiency in term of makespan, resource utilization metrics for set of tasks.

1. Introduction

In the last two decades, the high speed low-cost computers production led to boost the processing power through networks; this fact helps to grow in distribution of distributed systems to contain and tap these resources [1,2].

A heterogeneous computing (HC) referring to set of heterogeneous elements called computers, resources, processors or machines, these resources connected to each other by a network [2]. Figure 1 shows the heterogeneous computing environment.

Cloud computing and Grid computing are famous examples of distributed computing resources working over loosely coupled virtual machines, created by connecting many heterogeneous computing resources with different characteristics. Where both cloud and infrastructures provide an easy low cost solutions to access for distributed computing resources distributed over the globe to solve large problems [3,4].

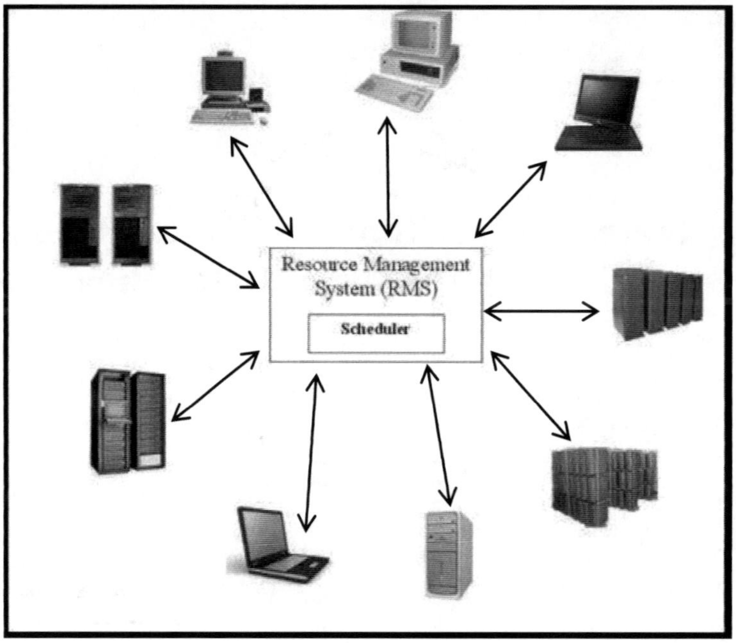

Figure 1. HC environment

The main issue when using a distributed computing systems environment is finding an efficient way for task scheduling to satisfy the user needs of fast computing power, where tasks should be assigned to the resources in a clever way [2].

In this research, we aim to propose an efficient solution for Heterogeneous Computing Problem (HCSP) and evaluate this method to ensure its efficiency.

2. Related Works

Tasks and machines in general are heterogeneous with different capabilities and different properties. The large scale of resource and big number of tasks can lead to system degrading; this fact motivated the researcher to do many researches on scheduling methods for heterogeneous computing systems. Resource heterogeneity, tasks heterogeneity and task dependencies were the main issues to deal with during conducting the research [5].

HCSP attracted many researchers, where many techniques used to solve this problem such as heuristic and metaheuristic techniques [5,6].

Min-Min algorithm introduced by [7] and [8]. Min-Min schedule the task depends on earliest finishing time (EFT) value for each machine. Despite its good results, scheduling in Min-Min seems bad idea when the number of short tasks is much higher than the number of long tasks where there is no many options for scheduling [9,**10**].

Authors in [11] and [12] proposed load balancing algorithm. Min-mean algorithm derived from Min-Min algorithm. Where this algorithm has two steps, in first step, Min-Min algorithm was used. In the second step, mean execution time (meanCT) computed and compared with all machines' finish time, if the machine with EFT less than meanCT, then the task with maximum EFT value will be scheduled, otherwise, the task with minimum EFT will be scheduled. The algorithm showed better results comparing to min-min algorithm in term of total execution time. But the problem of having majority of loaded machines was not solved because the algorithm depends on Min-Min algorithm in its first stage.

Unlike Min-Min which select depending on earliest finishing time, the MaxMin in [6,8,13] choose tasks to be scheduled depending on maximum EFT value[14]. Another problem appeared when the number of long tasks is much higher than the short tasks, where it might cause makespan to degrade less than Min-Min [10].

An improved algorithm Max-Average was developed by [15] to improve Max-Min algorithms by calculating the average completion time for all machines then assign the maximum completion time task to minimum estimated completion time resource. The new algorithm shows a better result comparing to Max-Min algorithm in term of makespan.

Sufferage discussed by [13] and [8] , where the tasks in Sufferage algorithm might suffer if the task does not schedule to the machine which can execute it faster. The suffer value compute by finding the difference between the machine with earliest finish time (EFT) and the machine with second earliest finish time. After calculating the suffer value for all tasks, the task with highest suffer value will be assigned to the machine with lowest completion time. The main drawback of this algorithm is calculating the suffer value in each iteration which cause more time to be lost.

LSufferage proposed by [1] is an enhancement of Sufferage algorithm trying to overcome its drawback by adding the priority list to store the priority of execution for each task, where each machine has a list of priorities. The priorities of tasks which run on faster machine are higher than other machines. The results of this algorithm didn't show better results in term of makespan comparing to other heuristics in the paper.

A heuristic algorithm with load balancing mechanism named penalty based (PB) were proposed by [16] . The tasks in this algorithm assigned to machine with minimum execution time in first phase. Second phase did the load balancing by moving the task from the most loaded machine to less loaded machine, task penalty in this case must be minimum.

TPB is an enhanced algorithm; this algorithm is an enhancement of PB algorithm proposed by [1]. The idea of moving the task from the most loaded machine to least loaded machine has been extended. Tenacious penalty based algorithm (TPB) move the task from the most loaded machine to any other machine by checking with other machines after sorting them depending on load in ascending order. The authors compared between the new two proposed algorithms (LSufferage and TPB) with some other heuristics. The results showed that their new algorithm TPB is better in term of makespan.

Even TPB showed excellent results in term of makespan, researchers always aim to reach optimality. In our research, we aim to reach an optimal solution to execute tasks faster and improve the utilization.

3. Experiment

Heterogeneous Computing Scheduling Problem (HCSP) become very important issue, especially with the increasing in use distributed computing with heterogeneous resources [17].

Estimated Tasks Completion (ETC) matrices proposed by [6,18] used to test and evaluate our proposed algorithm. ETC matrix contains the expected execution time for the task on each resource. Each element in matrix ETC[t,p] represents the execution time for task t in resource p, assuming that the transfer time of the task included. These values assume to be known a priori approaches that can be used. This estimated values based on analytical benchmarking and profiling [6]. With twelve datasets proposed by [6] which consider all heterogeneity and consistency for both tasks and machines, these testbeds become as one of main standards benchmarks which can evaluate and solve heterogeneous computing scheduling problem (HCSP). Table 1 below showing the heterogeneity levels for datasets used in our experiments

Table 1. heterogeneity and consistency for ETC model

Heterogeneity		Consistency		
Task	Resource	Consistent	Inconsistent	Semi- consistent
High	High	u_c_hihi	u_i_hihi	u_s_hihi
High	Low	u_c_hilo	u_i_hilo	u_s_hilo
Low	High	u_c_lohi	u_i_lohi	u_s_lohi
Low	Low	u_c_lolo	u_i_lolo	u_s_lolo

ETC model considered as one of most important computational models, work to solve scheduling problems as [2,5] mentioned. Full details, datasets and some experimental results can be downloaded from [19].

For this purpose, HCSP mathematical model developed by [1] to deal with the ETC matrices has been used. This model designed to achieve a better understanding for problem, and better scheduling results as well by concerning the makespan value minimization as a main objective.

4. The proposed Algorithm

Our proposed algorithm inspired by TPB algorithm proposed by [1] which has very efficient executing time.

Tenacious penalty based algorithm (TPB) proposed has two phases. In first phase, TPB sort the machines depending on their load in ascending order; assign the tasks to earliest finishing time machine. In the second phase, the tasks move from the most loaded machine to any other machine according to penalty value by checking with other machines after sorting them depending on load in an ascending order. Table 2 below shows the definition of notations used in pseudo code for both Algorithm 1 and Algorithm 2.

Table 2. Definition of notations

Notation	Definition
comp_a	Completion time of pa
cb	Completion time of pb
ETC	Estimated Completion Time
m_penalty	Minimum penalty
new_cb	New completion time for the selected machine
new-comp_a	new completion time for the Maximum loaded machine
new_ms	New calculated makespan
P	Set of all machines
pa	Most loaded machine
pb	Selected machine
pl	Set of machines sorted in ascending order
rep	Repetitions
s_proc	Selected machine
s_task	Selected task to be transferred from pa
s_t_1	Selected task from pa to be exchanged with s_t_2 from pb
s_t_2	Selected task from pb to be exchanged with s_t_1 from pa
T	Set of all tasks

Algorithm 1 Tenacious penalty based (TPB)

∀ t where t∈T , Schedule t to the machine which can execute the tasks faster
rep = 0

```
while rep < T
    rep = rep + 1
    m_penalty = +∞
    s_task = −1

    for all pb ∈ pl excluding pa do
        for all t ∈ tasks scheduled in pa do
            if cb +ETC(t, pb) < comp_a then
                penalty = (ETC(t,pb)-ETC(t,pa))/(ETC(t,pa))
                if penalty < m_penalty then
                    m_penalty = penalty
                    s_task = t
                    s_proc = pb
                end if
            end if
        end for
        if s_task ≠ −1 then
            Transfer s_task from pa to s_proc
            Continue the while loop
        end if
    end for
    if s_task = −1 then
        return
    end if
end while
```

The need for fast task execution time in distributed computing systems increased rapidly. This need was a motivation to propose an efficient algorithm to reduce the execution time and increase the utilization between the resources. In this section, a proposed algorithm Two Stages Tasks transfer (TSTT) presented. From its name, we can conclude that the proposed algorithm has two stages. The first stage adopted from Tenacious Penalty Based algorithm (TPB) algorithm proposed by [1], where the scheduler tries to reduce the makespan by transferring the task from most loaded processer to any other machine. The task and its receiver machine were chosen according to a penalty equation which can lead to reduce the makespan. The proposed algorithm TSTT extends the idea of TPB by adding another step where tasks can be exchanged between the two chosen machines depending on a comparison value between the estimated makespan of the chosen machine and the maximum loaded machine.

Algorithm 2 Two Stages Tasks Transfer (TSTT)

```
Stage 1: is equivalent to TPB algorithm (Algorithm 1)
Stage 2: Tasks exchange stage:

For all t1∈ T where t1 scheduled in pa
    For all p in P except the pa {
        Selected machine (pb)
        cb → (selected machine completion time)
        For all tasks t2∈ T where t2 scheduled in pb
            new-comp_a → (new completion time for the Maximum loaded machine)
            new-comp_a= comp_a- ETC(t1, pa)+ETC(t2, pa)
            new_cb = cb – ETC(t2,pb) + ETC(t1,pb)
```

```
    if (ca_new < comp_a && cb_new < ca) then
        new_ms= max(ca_new, cb_new); // new makespan
        If (new_ms<ca)  → s_t_1=t1
                        → s_t_2=t2
                        → s_proc=pb
          End if
        End if
    End for
      If tasks selected → transfer s_t_1 from pa to s_proc
                        → transfer s_t_2 from s_proc to pa
      End if
End for
```

The pseudo code presented in Algorithm 2 represented the details of the proposed algorithm. By checking if the task exchange between the maximum loaded machine *pa* and the selected machine pb will lead to reduce the makespan, then the scheduler will make a decision to do a task exchange. Otherwise the scheduler will keep searching till find a machine with the task which lead to decrease the makespan. This algorithm set reducing the total execution time (makespan) as a goal.

5. Results and discussion

In this paper, two important performance metrics have been used to show the efficiency of state of the art algorithms in term of makespan and resource utilization. The proposed algorithm will be compared with well-known heuristics like Sufferage, penalty based and Tenacious penalty based algorithms.

5.1. Parameters

The parameter settings performed depending on the twelve problem datasets [19] were used. In this experiment, datasets tested 512 X 16. This indicates 512 tasks and 16 machines. Java based simulation has been used on Intel (R) core(TM) i7 -6500U CPU @2.5 GHz, 2.9 GHz to run this simulation. The research has been done using the parameters shown in table 3 below:

Table 3. Parameters used in the implementation

Tasks	512
Machines	16
Task heterogeneity	High, low
Machine heterogeneity	High, low
Machine consistency	Consistent, inconsistent, semi-consistent
Datasets	u_c_hihi, u_c_lohi, u_c_hilo, u_c_lolo, u_i_hihi, u_i_lohi, u_i_hilo, u_i_lolo, u_s_hihi, u_s_lohi, u_s_hilo, u_s_lolo

5.2. *Performance metrics*

In this paper, two performance metrics has been used to show the efficiency of our proposed algorithm in term of makespan, resource utilization. The proposed algorithm compared with well-known heuristics like Sufferage, Penalty Based, and Tenacious penalty based algorithms.

5.2.1. Makespan

Makespan is the time takes the scheduler to schedule all tasks in its suitable resources, where the makespan is the main objective in scheduling process [20]. Makespan equation (1) listed below:

Makespan= max {completion[t], where t ∈ T } … (1)

In our experiment we used the lower bound (LB) values to represent the optimal execution time to solve the scheduling problem. These values are unrealistic values, because it is assuming that the system can't be interrupted under any circumstances like some physical or even task preemption. These values calculated using linear programming relaxation assuming every task in HCSP instances always executed on fastest machine (2). The LB values for all the twelve datasets published in [19].

The gap between the optimal solution and the current solution for the heterogeneous computing scheduling problem (GAP) presented in equation (2) were introduced by [21].

$$\text{GAP}=\frac{Current\ makespan-Lower\ bound\ value}{Lower\ bound\ value} \qquad (2)$$

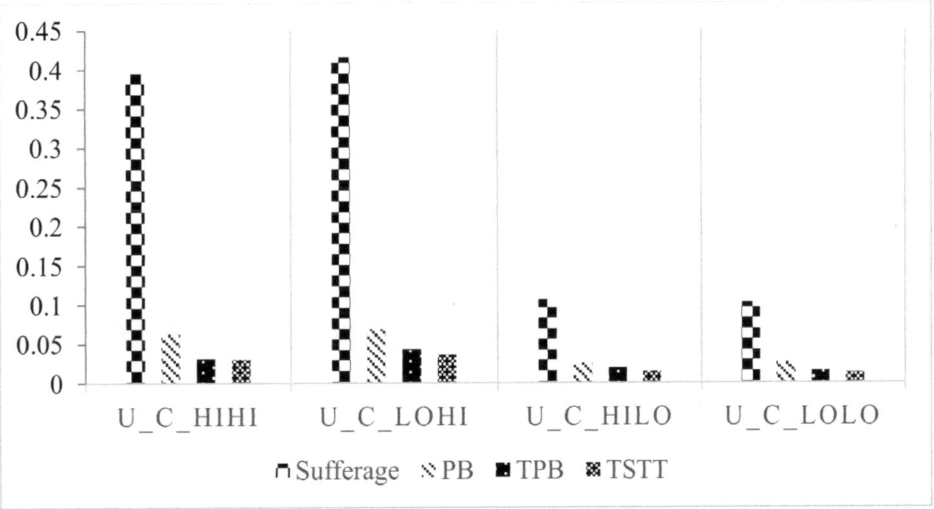

Figure2. The gap between calculated makespan and lower bound for the consistent datasets (u_c_hihi, u_c_loho, u_c_hilo, u_c_lolo).

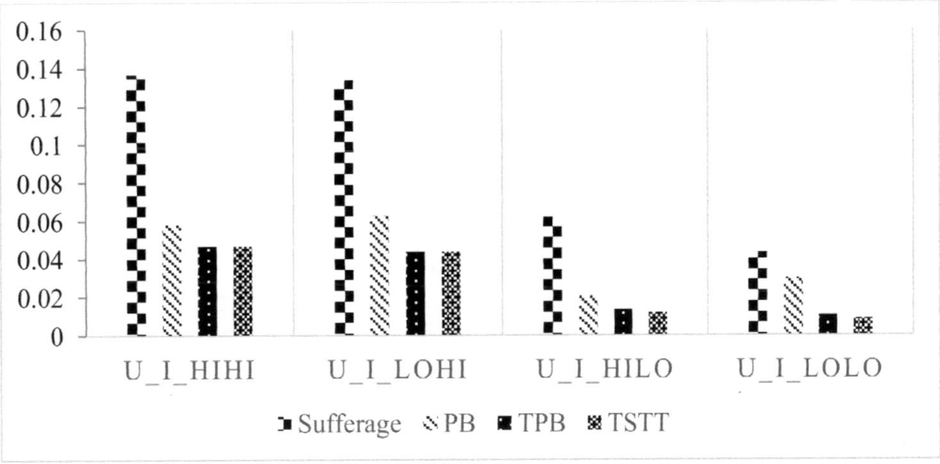

Figure 3. The gap between calculated makespan and lower bound for the inconsistent datasets (u_i_hihi, u_i_lohi, u_i_hilo, u_i_lolo).

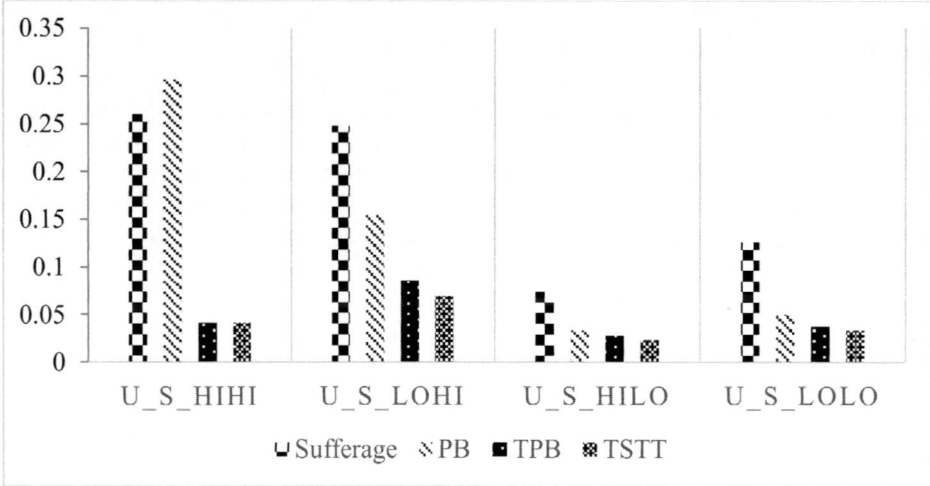

Figure 4. The gap between calculated makespan and lower bound for the semi-consistent datasets (u_s_hihi, u_s_loho, u_s_hilo, u_s_lolo).

The figures 2, 3, and 4 showed graphical summaries of the makespan gap from the optimal solution. The datasets categorized according to the consistency into consistent, inconsistent and semi-consistent. Figure 2 represents the consistence datasets, while figure 3 represents the inconsistent dataset, and figure 4 represents the semi-consistent one.

From the result shown above in figures 2, 3 and 4, we can observe that the bar of (TSTT) has a least value among all heuristics compared with it (Sufferage, PB and TPB). This means that the proposed algorithm is closer to optimality comparing to other examined algorithms.

5.2.2. Resource_Utilization

Resource utilization "is one of the main characteristics in heterogeneous computing. Achieving high resource utilization is the goal for researchers to gain benefits from idle resources. We can also say it is a quality of service metric. Resource utilization can be defined as the average utilization of the resources". Taking full advantage of resources is an important objective in heterogeneous computing scheduling. The importance of the resource utilizations because of the economic effect [5].

$$\text{Resource utilization} = \frac{\sum Completion\ time\ (p)}{makespan \times np} \qquad (3)$$

Where p ϵ P (Set of all the machines), and np= number of machines.

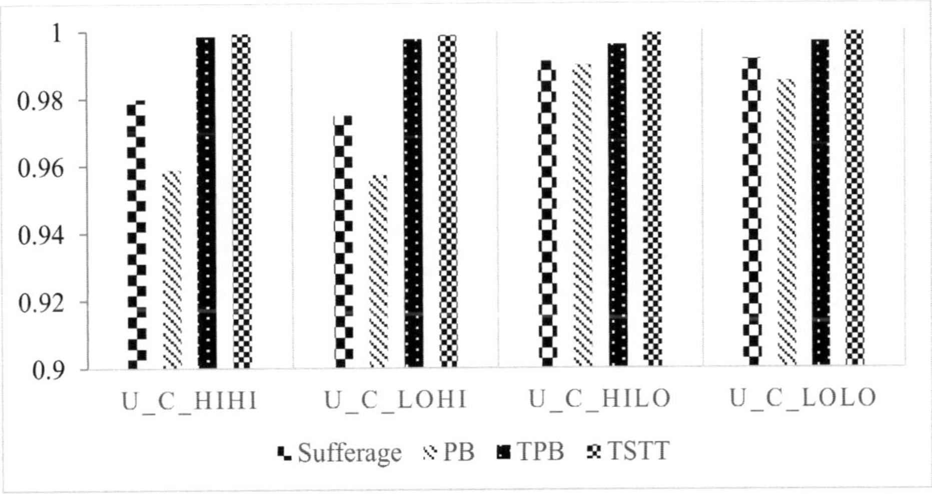

Figure 5. The resource utilization for the tested heuristics using the consistent datasets (u_c_hihi, u_c_loho, u_c_hilo, u_c_lolo).

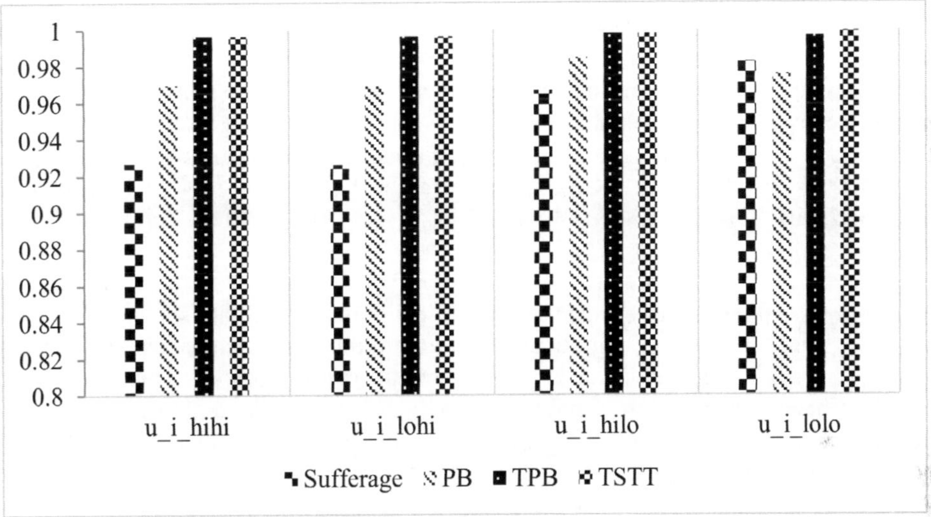

Figure 6. The resource utilization for the tested heuristics using the inconsistent datasets (u_i_hihi, u_i_loho, u_i_hilo, u_i_lolo).

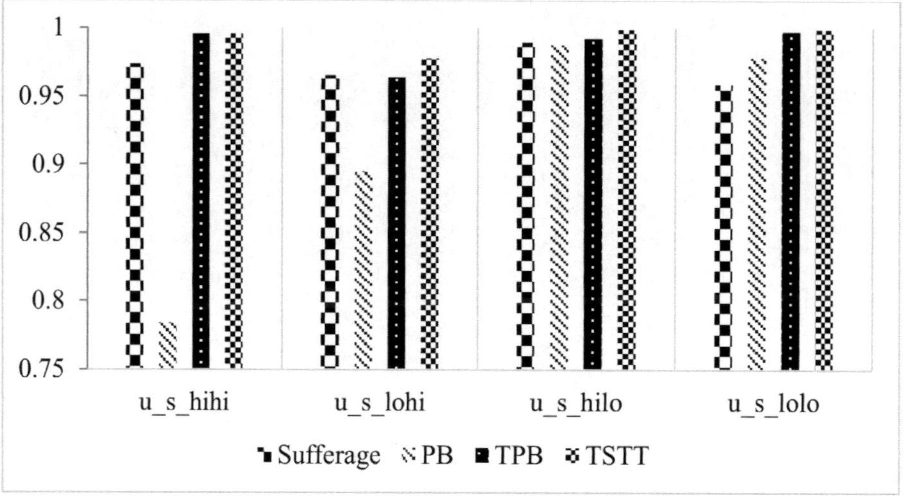

Figure7. The resource utilization for the tested heuristics using the semi-consistent datasets (u_s_hihi, u_s_loho, u_s_hilo, u_s_lolo).

The figures 5, 6 and 7 summarized the resource utilization results of the tested algorithms for all heterogeneity and consistency prospects. The datasets categorized according to the consistency into consistent, inconsistent and semi-consistent. Figure 5 represent the consistence datasets, while figure 6 represent the inconsistent dataset, and figure 7 represent the semi-consistent one. As shown in these figures we can observe that the orange color bar (TSTT) has a better utilization comparing to the other tested heuristics, where the result for the proposed algorithm is closer to the maximum utilization (closest to 100% utilization).

However, these results have not previously been discussed. Figure 2 showed the gap of produced makespan from the optimal makespan in consistent datasets. The results showed that the proposed algorithm always has a better execution time comparing to other heuristics; this might be because the new algorithm have a better load balancing for the reason that the proposed algorithm always make sure to produce better makespan by transferring and exchanging tasks to get better results in term of execution time.

Figures 3 and 4 presented the gap of produced makespan from the optimal makespan in inconsistent and semi-consistent datasets, the results showed that the proposed algorithm always produce better results from other examined heuristics, but in some datasets, such as u_i_hihi, u_i_lohi and u_s_hihi, this algorithm share the best results with TPB algorithm. These results may be explained by the fact that the TSTT is the proposed enhancement of TPB algorithm done by [1], where a new stage has been added to enhance the algorithm and produce better makespan by exchanging the tasks between the resources; This gives a reason where this result might be generated because there are no tasks have been exchanged because, the condition of tasks exchange in the second phase have not met.

We can notice that the same datasets (u_i_hihi, u_i_lohi and u_s_hihi) in resource utilization figures (5, 6 and 7); the proposed method TSTT and TPB algorithm got same values, while in other nine datasets produce better values. This might happen as we mentioned previously because of there are no tasks have been exchanged because the condition of task exchange in the second phase have not met, where the results of resource utilization influenced by makespan value [20].

6. Conclusion

The scheduling in heterogeneous computing system is critical problem; especially with the increasing in heterogeneous resources usage in different heterogeneous environments like grid computing and cloud computing with increasing in the tasks submitted to these resources. This fact lead to increases the need to provide an efficient task scheduling method which can be reduce tasks execution time, and utilization between resources to achieve fast services for the system clients and optimal utilization between resources.

An enhanced algorithm proposed in this paper used a load balancing technique to provide better results in term of makespan and resource utilization. Two stages tasks transfer mechanism used to achieve better load balancing by moving the tasks from most loaded machine to any other machine in a way where make sure that the new makespan always lesser than the previous one. The results showed that the proposed method provide better results in term of makespan, resource utilization than other well-known scheduling algorithms.

For future work, we plan to consider other specification, such as: tasks deadline, and task preemption in our research.

References

[1] Gogos C, Valouxis C, Alefragis P, Goulas G, Voros N, Housos E. 2016 Scheduling independent tasks on heterogeneous processors using heuristics and Column Pricing. *Futur Gener Comput Syst*. **60**:48–66.

[2] Nesmachnow S. 2013 Parallel multiobjective evolutionary algorithms for batch scheduling in heterogeneous computing and grid systems. *Comput Optim Appl*. **55**(2):515–44.

[3] Foster I, Kesselman C. 2004 The grid 2: Blueprint for a new computing infrastructure [Internet]. Morgan Kauffman. p. 748. Available from: http://scholar.google.com/scholar?hl=en&btnG=Search&q=intitle:The+GRID+2#4%5Cnhttp://books.google.com/books?hl=en&lr=&id=0I5gm6o3vrMC&oi=fnd&pg=PP2&dq=The+grid+2:+Blueprint+for+a+new+computing+infrastructure&ots=hoHjwTKvrd&sig=KZWHe1mc-KqVg7_nkkEpfyhCb6U

[4] Panda SK, Gupta I, Jana PK. 2015 Allocation-aware task scheduling for heterogeneous multi-cloud systems. *Procedia Computer Science*. p. 176–84.

[5] Xhafa F, Abraham A. 2008 Meta-heuristics for grid scheduling problems. Stud Comput Intell. **146**:1–37.

[6] Braun TD, Siegel HJ, Beck N, Bölöni LL, Maheswaran M, Reuther AI, et al. 2001 A Comparison of Eleven Static Heuristics for Mapping a Class of Independent Tasks onto Heterogeneous Distributed Computing Systems. *J Parallel Distrib Comput* [Internet]. **61**(6):810–37. Available from: http://dx.doi.org/10.1006/jpdc.2000.1714

[7] Ibarra OH, Kim CE. 1977 Heuristic Algorithms for Scheduling Independent Tasks on Nonidentical Processors. *J ACM*. **24**(2):280–9.

[8] Casanova H, Legrand A, Zagorodnov D, Berman F. 2000 Heuristics for scheduling parameter sweep applications in grid environments. In: *Proceedings 9th Heterogeneous Computing Workshop (HCW 2000) (Cat NoPR00556)* [Internet]. p. 349–63. Available from: http://ieeexplore.ieee.org/lpdocs/epic03/wrapper.htm?arnumber=843757

[9] Anousha S, Ahmadi M. 2013 An improved Min-Min task scheduling algorithm in grid computing. *Lecture Notes in Computer Science (including subseries Lecture Notes in Artificial Intelligence and Lecture Notes in Bioinformatics)*. p. 103–13.

[10] Tabak E, Cambazoglu B, Aykanat C. 2013 Improving the Performance of Independent Task Assignment Heuristics Minmin, Maxmin and Sufferage. *IEEE Trans parallel Distrib Syst*. **25**(5):1–14.

[11] Kamalam G, Muralibhaskaran V. 2010 A new heuristic approach: Min-Mean algorithm for scheduling meta-tasks on heterogeneous computing systems. *Int J Comput Sci Netw Secur*. **10**: 24–31.

[12] Kamalam G, Bhaskaran V. 2012 New enhanced heuristic min-mean scheduling algorithm for scheduling meta-tasks on heterogeneous grid environment. *Eur J Sci Res*. **70**(3):423–30.

[13] Maheswaran M, Ali S, Siegel HJ, Hensgen D, Freund RF. 1999 Dynamic Mapping of a Class of Independent Tasks onto Heterogeneous Computing Systems. *J Parallel Distrib Comput*. **59**(2):107–31.

[14] Chang RS, Lin CY, Lin CF. 2012 An Adaptive Scoring Job Scheduling algorithm for grid computing. *Inf Sci (Ny)*. **207**:79–89.

[15] Maipan-uku J, Muhammed A, Abdullah A, Hussin M. 2016 Max-Average: An Extended Max-Min Scheduling Algorithm for Grid Computing Environtment. *J Telecommun Electron Comput Eng.* **8**(6):43–7.

[16] Chaturvedi AK, Sahu R. 2011 New Heuristic for Scheduling of Independent Tasks in Computational Grid. *Int J Grid Distrib Comput*. **4**(3):25–36.

[17] Nesmachnow S, Canabé M. 2011 GPU implementations of scheduling heuristics for heterogeneous computing environments. *XVII Congreso Argentino de Ciencias de la Computación.*

[18] Ali S, Siegel HJ, Maheswaran M, Hensgen D. 2000 Task execution time modeling for heterogeneous computing systems. *Proceedings 9th Heterogeneous Computing Workshop (HCW 2000) (Cat NoPR00556)*. p. 185–99.

[19] CECAL: HCSP- Heterogeneous Computing Scheduling Problem [Internet]. CECAL. 1993. Available from: https://www.fing.edu.uy/inco/grupos/cecal/hpc/HCSP/

[20] Gogos C, Valouxis C, Alefragis P, Xanthopoulos I, Housos E. 2016 Scheduling independent tasks on heterogeneous computing systems by optimizing various objectives. *Proceedings of the 11th International Confenference on Practice and Theory of Automated Timetabling (PATAT-2016)*. p. 149–61.

[21] Nesmachnow S, Cancela H, Alba E. 2012 A parallel micro evolutionary algorithm for heterogeneous computing and grid scheduling. *Appl Soft Comput J.* **12**(2):626–39.

Institute of Physics
Dirac House, Temple Back
Bristol BS1 6BE UK

ISSN: 1742-6588
ISBN 978-1-5108-6401-6

CURRAN ASSOCIATES INC.
proceedings.com

International Conference on Building Materials and Materials Engineering (ICBMM 2017)

IOP Conference Series: Materials Science and Engineering
Volume 264

Lyon, France
21 – 23 September 2017

IOP
Institute of Physics